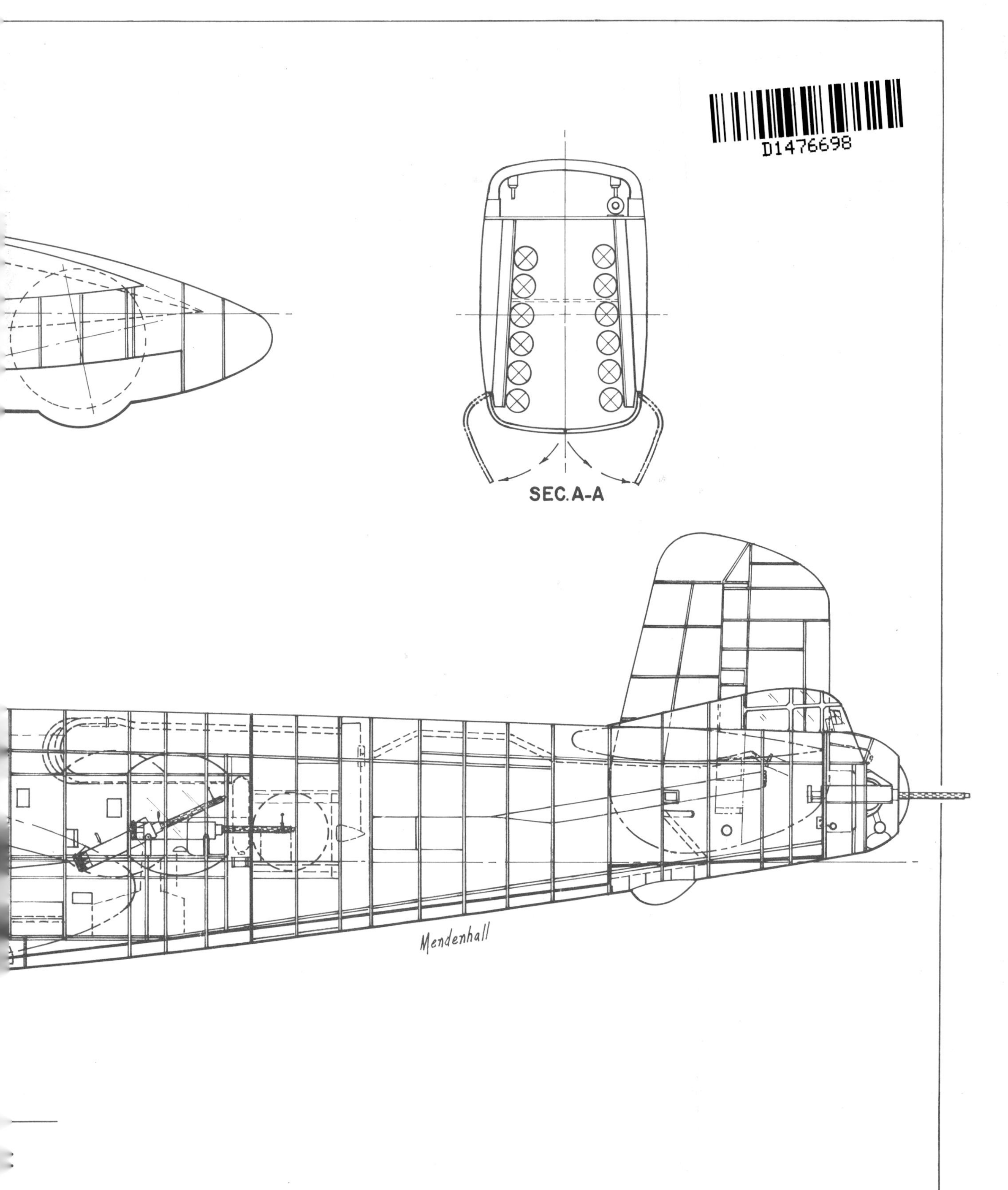

REF: NORTH AMERICAN AVIATION
KANSAS CITY DIVISION
DWG. NO. 108-0003

Ex Libris
Burl & Carol
Estes

DEADLY DUO

The B-25 and B-26 in WW-II

CHARLES A. MENDENHALL

For Diane

Books by Charles A. Mendenhall
The Air Racer
The Gee Bee Racers: A Legacy of Speed
Deadly Duo: The B25 and B26 in World War II

First Printing—June 1981

ISBN 0-933424-22-1
LC 80-52625

Published by: Specialty Press Publishers and Wholesalers, Inc.
Osceola, WI 54020

Distributed in the U.K. by
Midland Counties Publications
Earl Shilton, Leicester
England

Contents

Drawings

INTRODUCTION

THE DEVELOPMENTAL IMPACT of the Norden bombsight had a momentous effect on military thinking during the 1930's. The curious, super-secret instrument was first rumored, then widely publicized, as being capable of dropping a bomb in a pickle barrel from thirty thousand feet. When asked if this was really true the inventor countered, "Which pickle would you like to hit?" With that heady capability in mind, the Army brass could envision a fleet of Norden-equipped strategic bombers cruising high in the stratosphere totally impervious to attacking fighters and antiaircraft fire. The aerial armada could attack an enemy across vast distances with deadly accuracy. It was great—but to accomplish this feat, it would require big aircraft. And they were thinking big, too. As early as 1934, the Army Air Corps (AAC) issued a requirement for a bomber with a range of 5,000 miles capable of cruising along at 200 miles per hour carrying a ton of bombs. Labeled "Project A," such aircraft could even protect the outlying U.S. territories of Alaska, Hawaii and Panama.

Carl L. Norden, Theodore H. Barth and Frederick I. Entwistle, assistant research chief of the Navy Bureau of Ordnance, had begun development of the Norden sight in 1928. When perfected, it would become a sophisticated beginning to black box warfare. The Norden sight was linked to the bomber's auto-pilot in such a manner that the bombardier actually flew the aircraft during the bombing run. It automatically calculated the target's movement, wind drift, the bomber's forward movement, and it made the necessary compensations to assure a direct hit on the target. It even *released* the bombs when the aircraft was positioned properly. With bombs away, the pilot took over control again.

The secrets of the Norden sight were so carefully guarded that all bombardiers carried a Colt .45 automatic pistol as they packed their mechanical/optical instrument in a leather satchel to the plane and back. Their orders were to guard the sight with their lives and shoot to kill if need be.

With the 1934 requirement for a big bomber, and the Norden bombsight available, it was not long before the always agile Boeing Airplane Company obtained a contract for their proposed Model 294, or XB-15. It was to be the largest military aircraft ever built; however, it did not fly until October 15, 1937. The large, thick wings spanned 149 feet and its fuselage length was 87 feet 7 inches. Fully loaded, it weighed in at 70,706 pounds, with four 2,000-pound bombs accounting for some of that heft.

The XB-15 was destined to be a one-of-a-kind, due to an unfortunate lack of suitable engines. The problem was not in building an aircraft that large, but in powering it. The best engines available were Pratt & Whitney's Twin Wasps rated at 850 horsepower—only 1,000 horsepower at takeoff, rather weak for that size of aircraft. The XB-15 later became the XC-105 and was used as a cargo carrier during World War II. It was then written off and scrapped near the war's end.

Early in the XB-15 program, however, Boeing recognized the anemic powerplant problem and backed away a couple of notches in airframe size. They began work on their Model 299, later to be designated the XB-17. Its shorter wing span was 103 feet 9 inches and the length was 68 feet 4 inches. The prototype was quickly designed and built. It first flew July 28, 1935, and after a few false starts, eventually became one of our standard heavy bombers of the late 1930's and World War II.

Another interesting exercise of Project A was the gargantuan Douglas XB-19. Designed and constructed late in the thirties, it first flew July 27, 1941. With a huge 212-foot wing span, it had an immense length of 132 feet 4 inches. Once again, the airframe manufacturers had outstripped the ability of the powerplant builders to keep up. Even with four 2,000-horsepower Wright Cyclones, the ship was underpowered. Like the XB-15, a single copy was built and used only for a test vehicle until it was scrapped in 1949.

The design of the Consolidated B-24 "Liberator" was a more formidable high-flying stratospheric bomber attempt. Its size and power were more in keeping with the successful B-17 "Flying Fortress" and it had the advantage of being able to use the lessons learned on the B-15, B-17 and B-19 aircraft. (A

B-16 heavy bomber designed by Glenn L. Martin was to have been powered by four engines but it never got off the drawing board.) The B-24 first flew December 29, 1939, only nine months after Consolidated was awarded a contract for its construction.

Thus, the Army Air Corps ended the thirties with two very satisfactory long-range, high-altitude heavy bombers—the B-17 and B-24—both to be used in conjunction with the Norden bombsight. In June 1937, Major General Frank M. Andrews predicted that in the future only four-engined bombers would be bought because "they are the basic element of air power." And so it was. Targets deep within enemy territories could be demolished at will, at least in theory. All that remained to be done was crank up the giant American industrial facilities and build the big bombers by the thousands.

Even the most diehard big-bomber proponents admitted, however, there *was* a need for light, tactical aircraft flying close support missions for ground troops. One development program for such aircraft had run parallel with the effort on the heavy bombers. The Douglas A-20 (DB-7), a smaller twin-engine design, was first flown as a prototype at the end of 1938. Forty-eight feet long with a span of 61 feet 4 inches, the *Havoc,* as it was known, fulfilled the Army's requirement for a light aircraft with a lot of punch close-in against enemy ground forces.

Prior to the AAC circular that had brought forth the DB-7, all attack bombers had been single-engine ships with very limited carrying capacity; for example, the Northrop A-17 and Curtiss A-18. The circular, which called for a light twin-engine bomber, also brought out contemporary prototypes such as the Stearman XA-21 (it lost out in the competition); and the Martin XA-22 that eventually became the "Maryland" lend-lease bomber that went to Great Britain in large quantities. Then there was the ill-fated North American NA-40-2 prototype which crashed even before being given an "A" dash number. (This latter aircraft, however, was far from counted out and a great deal more will be heard about it.)

Still other light attack bombers were used in service even though they were single-engined. Examples were the Douglas A-24 (the Navy's famous SBD "Dauntless" in olive drab); the Curtiss A-25 (the Navy's SB2C "Helldiver") and the North American A-36 (the P-51 "Mustang"). It is apparent that a number of light short-range tactical bombers had been designed and were available in addition to heavy long-range strategic bombers. It was then decided a third classification of bombers would be useful—the medium bomber—a sort of in between size compared to the heavy and light machines.

The new medium aircraft would carry about half the bomb and fuel loads of the heavies. The AAC Circular Proposal 39-640 was put out on March 11, 1939, for the new bombing aircraft. In answer, two designs emerged and went into production literally off the drawing board: the North American B-25 and the Martin B-26.

In May of 1940, President Franklin D. Roosevelt gave the Allies a ray of hope for ultimate victory. He called for the production of 50,000 war planes in the United States, and the B-25 and B-26 would eventually be a significant part of that effort.

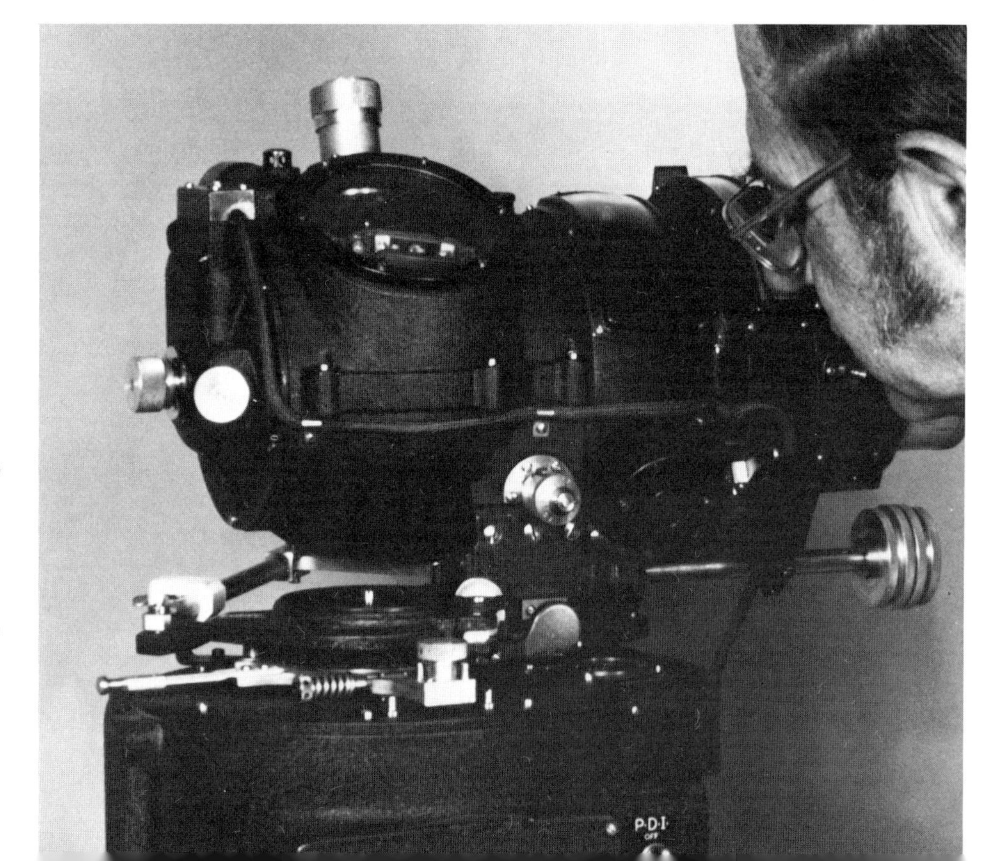

The famous Norden bombsight that, amongst other triumphs, eventually led to the design and building of the B-25 and B-26 medium bombers. It was sometimes known as "The Blue Ox" in the China theater of operations and "The Football" in Europe. (Norden Systems)

1

CANNON CARRIER EXTRAORDINARY

July 28, 1943—Port Moresby—(INS)—Airborne 75 millimeter cannon fire erupted today in the Bismark Sea off the coast of New Britain. For the first time, an American B-25 from the Third Attack Group based at Three-Mile Airstrip near Port Moresby sank a Japanese destroyer using cannon fire alone. The pilot of the aircraft was Air Corps Col. Paul I. Gunn. In other areas of the Solomons. . . .

THIS BRIEF NEWS STORY was the beginning of another facet in the colorful story of the North American B-25 "Mitchell" bomber. The first bomber over Tokyo, a stalwart in the North African and Mediterranean campaigns and a steady deterrent in the cruel cold of the Aleutian Islands, it had once more risen to noteworthy acclaim. The first cannon-fitted B-25 was a bit of a lash-up, but was so effective that North American quickly came out with a real gunbus—the B-25G. Then came the B-25H which carried fourteen .50-caliber machine guns *plus* a 75-millimeter cannon. Superplane—super killer!

Large oil storage tanks, such as the ones in the Lanywa oil fields in Burma, soon fell victim to the hole punching power of airborne big-bore shells. An oil storage tank, punctured and set afire by one or two of the big bangers, literally erupted in a fiery blast with a column of thick black smoke visible for miles, depriving the enemy of massive amounts of fuel.

Other targets, far afield from the U.S. Army's artillery but in need of its attention, soon were being blown to pieces by the new cannon/airplane duo. The large 75-millimeter shells, nearly three inches in diameter, were takeoffs of the ones that had pounded the German Army into submission in World War I. At that time, they were fired from standard field cannons of the French and United States armies. So famous were the 75's contributions to the first Allied victory that when General Motors in 1942 geared up for war production of its own fighter plane design it asked that the "P" numbers be jumped ahead to P-75. (The abortive project, called "Eagle," was soon dropped—a $50 million error in judgment.)

The massive airborne cannon was certainly something new. The 75, when coupled with a Mitchell's forward-firing machine guns, combined the wallop of a naval destroyer and the speed and maneuverability of a light bomber. The B-25G and B-25H models were first-class, destructive, lethal weapons.

The cannon shell's projectile was 2.9527 inches in diameter and nearly 12 inches long. It was filled with TNT and timed for explosion by a clockwork mechanical time fuse. (This fuse, incidentally, was just like the works of Grandpa's old alarm clock—all gears and springs, except instead of the 6:00 A.M. chime going off, the firing pin went into the primer and the shell exploded.) The overall length of the cannon shell was about two feet and it weighed nearly twenty pounds. Since there were twenty-one of them racked up in the aircraft, they totaled out to over four hundred pounds of high-explosive material stowed behind the pilot's seat! Another comparison shows the shell weighed more than seventy-eight rounds of .50-caliber machine gun bullets. It had an accurate range of over 10,000 feet—about two miles. With a direct hit, it could turn a light tank inside out. Conversely, with a hit of any kind of bullet in the aircraft's shell magazine area, there would be instant blowup of the crew and flying machine. To help prevent this, the magazine rack and the crew were armor-plated with 3/8-inch plates. Most crewmen would have preferred something more like a foot thick.

The North American XB-25G-NA (41-13296) started out as a B-25C but was converted to house a 75 mm cannon in a new, somewhat shortened nose. (USAF)

This B-25G-10-NA (42-65128) was in the last build group of the 306 G model cannon carriers completed. Silhouetted against the sky, the short pug nose is most evident. (USAF)

Another view of 42-65128 shows the twin .50-caliber guns above the cannon to aid in sighting, the front gunsight post atop the nose, and the armor plating covering a portion of the pilot's quarter panel window. (USAF)

Close-up of the massive 75 mm cannon. Also note heavy bevel-edged armor plate to protect pilot and shells from enemy gunfire. The aptly named "Pistol Packin Momma" operated in the Southwest Pacific in 1944. (USAF)

A dummy tank goes up as a B-25G lets go with a round of its 75 at the Air Proving Grounds in Orlando, Florida, early in the war. (USAF)

Over southern California, this B-25G-10-NA (42-65199) is tested before being shipped out for combat. The twin .50 Bendix dorsal turret is located in the same position as on the B-25C and D models. (USAF)

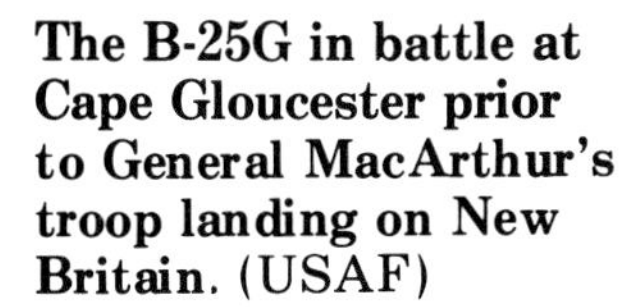

The B-25G in battle at Cape Gloucester prior to General MacArthur's troop landing on New Britain. (USAF)

The Japanese lie low in their entrenchments as a B-25C/D works them over with its four field-fitted nose .50's—the face on the nose is enough to scare the devil out of anyone even without that big, heavy firepower over New Britain. Incidentally, the guns in the enemy emplacements are 75's. (USAF)

264758

Four interesting views of B-25G (42-64758) in early USAAF maritime scheme—olive drab top with white bottom, red mouth, white teeth, red lips and blood drips. It is perplexing that this aircraft has a Model C serial number. (USAF)

The cannon itself was a standard M-4 U.S. Army field gun, 9½ feet long and weighing in at nearly nine hundred pounds. The initial trial installation was in a Douglas B-18 "Bolo" bomber. This aircraft was a medium bomber derivative of the famous Douglas DC-3 transport and was somewhat larger than the B-25. Initial trials with the B-18 gun platform proved quite successful, so the operational B-25 was tried next.

The 75's recoil was 21 inches. The impact was absorbed to some extent by a hydraulic piston and spring mechanism mounted integrally atop the gun's breech. A loading tray extended behind the breech to allow the gunner to extract spent shells and slam home a new one, just like on the front in World War I. For the first installation a B-25C-1 (serial number 41-13296) was selected; it was redesignated XB-25G. The mounting position of the 75 was in the lower left-hand side of the fuselage in the bombardier's passageway, the barrel and breech running beneath the pilot's seat. The gun mount consisted of a heavy structural cradle that securely joined the cannon to the airplane. Eventually the cannon evolved into the T9E1 model with an automatic feed mechanism designed to increase the rate of fire. That effort was dropped, however, as lighter, more powerful rockets came into use before the automatic cannon could be used in combat.

Upon completion of the initial installation, the cannon prototype aircraft was flown to the Columbia, South Carolina, AFB for firing trials and it was not long before the flying 75 was booming away at off-shore targets near Myrtle Beach. So far so good. Except for the deafening blast in the cockpit, the acrid smell of cordite and the ship's lurch in reaction to the recoil, the tests were successful. One problem did pop up a little later in the tests—fractures occurred in the fuselage skin rivet joints near the barrel due to muzzle blast. This was quickly fixed by adding another three inches to the barrel length, thus getting the blast away from the fuselage.

A jump ahead in the B-25's development came with the H model, 1,000 of which were built. They contained many new and interesting improvements over the G's including a new tail turret. (USAF)

This side view of a B-25H-NA (43-4198) shows off the snub nose for the cannon and four .50's and the bay window waist guns with .50's. Also, as can be seen, the periscope-sighted ventral turret is deleted and the twin .50 dorsal turret is moved forward to add room for the footwork of the waist gunners. (USAF)

Lift off with 20-degree flaps for another mission with four package guns on fuselage sides, four .50's in nose above cannon and two in top turret. Now that's a gun bus. (USAF)

Two fixed .50-caliber machine guns were mounted in the nose above the cannon, each supplied with 400 rounds. The upper half of the aircraft's nose swung upward on hinges like the hood of a car, for servicing and replenishing the ammunition supply. Firing of this flying artillery was accomplished by the pilot aiming the whole aircraft, sighting through a gunsight and touching off the twin .50's to get the target lined up. Tests showed that three 75 shots —or even four if there was a particularly energetic navigator/gunner aboard—could be gotten off during a normal strafing run. The whole procedure was not easy, however. It meant holding a straight and level course, during which time one was a sitting duck for anti-aircraft and small arms ground fire. Too, on a few occasions the cannon breech itself exploded within the cockpit (a problem which was never remedied).

Paul Gunn, a colorful character, gaunt of frame and eagle-eyed, was forty-six when he began fighting the Japanese. He had served a twenty-year hitch in the Navy as an enlisted pilot and retired before the war broke out. With all that experience, he became general superintendent of flying for Philippine National Airways. It was a good job and he lived happily with his wife and four children in Manila. The Japanese attack changed that. Gunn was in Java when it happened. He tried rescuing his family by flying a Twin Beech to Zamboanga to pick them up. It didn't work. The enemy fighters shot him down during the long over-water flight. Fortunately, he was picked up and taken by a friendly patrol boat to Australia. Once there, he returned to uniform—this time an officer with the rank of captain in the U.S. Army Air Corps.

A vast knowledge of the Philippine archipelago, based on his recent airline work, made Gunn invaluable. By April 10, 1942, a long-range mission under the command of Brigadier General Ralph Royce was mounted from Melbourne to Luzon via the Del Monte Pineapple Plantation airstrip on Mindanao with a 1,900-mile first leg. The long range was possible for the fifteen B-25's involved because of additional bomb bay fuel tanks constructed under Gunn's direction. Though the Japanese Army was moving forward relentlessly through the Philippines, Gunn and his bombers nevertheless got off several raids against them on Luzon and the Visayans while based at the Del Monte strip. Finally, Mindanao came under attack and it was time for retreat back to Australia. Gunn flew out, his plane loaded with soldiers fleeing the oncoming enemy. His combat experience thus far showed up a flaw in the otherwise perfect B-25: Missions over the jungle were low-level attacks and he needed more forward fire power to really do a destructive job of strafing.

About this time Jack Fox, a North American technical representative, arrived in Australia and the two men hit it off at once. Fox was small, aggressive and inventive. Soon, he and "Pappy" had rigged up four .50's in the bombardier's glass nose, all controlled by the pilot. That *was* an improvement. Next, they added two .50-caliber package guns on either side of the fuselage, which provided a battery of eight .50-caliber guns firing forward instead of the single .30-caliber gun on the original B-25. If that worked—and it sure did—how about a cannon? Since a bombardier couldn't get into the nose anyway, with all the guns and ammunition up there, Fox and Gunn added a 20-millimeter cannon in the bombardier's crawlway. Raids by Gunn and associates around New Guinea raised general hell with Japanese freighters and barges by blowing holes through their thin sides and sinking them. And their effect on a Japanese invasion convoy caught in the Bismark Sea was nothing short of sensational.

Fox, meanwhile, had been keeping North American Aviation (NAA) President James H. "Dutch" Kindelberger appraised of the field modifications and that was when it was decided to build a production model that could carry the 75. The B-25's nose was shortened and four .50's were mounted in its now solid metal contours along with the cannon. It became the B-25G model and 406 units were built and sent to the Pacific.

Next came the B-25H. Its main contribution to even greater firepower was the movement of the Bendix top turret and its twin .50's forward to what *was*

Four different views of B-25H-NA (43-4110), the sixth H model built. These photos provide a fine overall view of an H's outside configur-ation. (USAF)

Tengchung, China, is hit by B-25H's from the 14th Air Force, as the long white plume of smoke from the town's center attests. (USAF)

Col. Philip Cochran's (Flip Corkin for "Terry and the Pirates" fans) Air Commandos had moved British troops behind the Japanese in Burma when this photo was taken. On March 18, 1944, this B-25H from the 1st Air Commando force attacks enemy stores and supply depots at Wunto, Burma. (USAF)

High above the South China countryside on June 3, 1945, this B-25H concentrates on retaking its old 14th Air Force base at Linchow. The base had been captured by the Japanese on November 7, 1944. (USAF)

Small fragmentation bombs drop from open bay of B-25H of 1st Air Commando Group, 10th Air Force on Simpi, Burma. (USAF)

A Japanese destroyer escort near Armoy, China, takes some roughing up by a B-25J. The aircraft is from the 345th Bomb Group "Air Apaches." The date was April 6, 1945. (USAF)

"Little Joe," a B-25G, skims along peacefully over a typical Pacific coral atoll. It looks as if the pilot has his elbow out the side window à la Sunday drivers. That's an H-type tail turret however—an unusual mix of models. (USAF)

This 75 mm automatic shell feeding mechanism for the T9E1 aircraft cannon now sits on display beside the B-25 at the Air Force Museum. While it would have increased the rate of fire over the manually loaded guns, it came too late as the lighter, more powerful rockets took over. (Mendenhall)

Smoke billows below from burning supply depots and road and rail junctions at Sagaing, across the Irrawaddy River from Mandalay. Damage was caused by more than fifty tons of incendiaries dropped by B-25H's of the 12th Bomb Group from 10,000 feet. (USAF)

Two rockets nestle in the cannon tube and one is mounted under the fuselage in this installation photographed at the armament lab, Air Technical Service Command Headquarters, Wright Field. (USAF)

the navigator's compartment. This created a total of ten forward-firing .50-caliber weapons plus the 75-millimeter cannon. Big waist gun bay windows were also added, allowing gunners a larger field of fire, and a new canopied twin .50 tail turret finished off the job. The B-25H was a veritable arsenal of automatic weaponry. In addition, the B-25 could still tote a bomb bay full of low-level parachute fragmentation bombs to flatten anything left of a target after its strafing barrage. One thousand of this model were produced and shipped out to the Pacific theater.

The G's worked over New Britain, lobbing 1,253 rounds of 75-millimeter shells at the enemy, plus an accompanying mega-thousand rounds of .50-caliber bullets during a short period of thirty-six days. Targets included anti-aircraft gun positions and all sorts of shipping and ammunition dumps. By the time the Marshall Islands campaign was underway in February 1944, targets worthy of a 75's attention had just about dried up.

With the nose sections of these aircraft filling with noxious fumes from the cannon breech, and the fact that the gunners had to literally flatten themselves against a bulkhead to avoid getting banged around by the cannon's recoil, the use of the weapon declined to just about zero. Throughout the Pacific the cannon was pulled and two to four more .50-caliber guns were added in its place. Some G and H models began to appear with as many as eight Brownings sprouting from their snub noses.

Only in a few other cases was such a heavy cannon ever successfully mounted on an airframe. The British Mosquito XVIII (approximately 27 built) was equipped with a 57-millimeter Molins cannon, one of which sank a U-boat near the French coast in March, 1944. The Italians experimented with a 105-millimeter gun in a Piaggio P 108A and Beechcraft fooled around with a 75-millimeter mounted in its XA-38 "Grizzly." Neither of these latter two programs developed into anything other than an experiment. Yes, the B-25 was the only cannon carrier extraordinary.

2

THE UNIDENTICAL TWINS

IT WASN'T UNUSUAL FOR B-17 crewmen to put down the B-24—that the angular Liberator was the box the sleek Flying Fortress came in. The square-lined B-25 offered the same logic train when compared to the hotshot B-26. The B-25 was functional, no-nonsense design engineering without frills. An opposite, the B-26 was sculptor's art reaching deep into the aeronautical engineer's bag of tricks. Both approaches turned out extremely successful. Each aircraft was claimed by admirers and fans to be the best medium bomber of World War II—and *that* argument will go on forever. The real winner was the U.S. Army Air Corps. It was blessed with two splendid designs which went into production concurrently, unidentical twins, whose trials and triumphs led to their ultimate success in helping destroy the Axis war machine. However, while they were indeed great aircraft, they were not without a worldwide peer group. Overseas, the Germans, British, Italians and Japanese had also recognized the need for medium bombers.

Amongst them was the German terror of the London blitz, the Heinkel He 111, a few feet larger in span and length than the American mediums. Powered by two Junkers Jumo 211F-2 engines of 1,340 horsepower, it could move along with a maximum speed of 258 miles per hour. It was well armed and it could dump one or two tons of bombs per plane on its victims. Though a little outdated (its prototype first flew in 1935), it continued to soldier competently on for the Germans until the end of World War II. Another Teutonic titan was the Junkers Ju 88, a battle tiger that first flew in December of 1936. Its splendid design was readily adaptable to new engines and constant updating for specific battlefield requirements. With a span and length about the size of the B-25, it had generally the same performance as its equally square-lined American cousin.

The British Vickers Wellington was a larger aircraft than the U.S. mediums but not large enough to be called a heavy in the sense of the Lancaster and Stirling four-engine bombers. Its rugged geodetic construction served it well, and while it whacked along at only 250 miles per hour tops it could lay down a whopping 6,000 pounds of bombardment on the enemy.

The Italians fielded the Savoia-Marchetti S.M.79 Sparviero, a tri-motor job about the size of the B-25. While nice-handling and a respectable combat plane, its career was aborted due to the conclusion of Italian combat efforts rather early in the war.

The Japanese hopped around the Pacific in their slick medium, the Mitsubishi Ki.21, Type 97, Sally. It served well on every Japanese front while toting a maximum bomb load of 2,200 pounds with a respectable top speed of 297 miles per hour.

These five aircraft of foreign design are only representative of the B-25/B-26 peer group. All belligerent countries had various aircraft designed and produced with the medium bomber mission in mind.

The Douglas B-18 "Bolo" had DC-2 wings, DC-3 tail surfaces and a fuselage all its own. The Army took first delivery in early 1937; 134 were built. (USAF)

The Douglas B-18A with extended nose for easier use by nose gunner and bombardier at the same time. 217 were bult. Twenty went to the RCAF and were known as "Digbys." (USAF)

The Douglas B-23 "Dragon" was still another bomber development of the DC-2/3 transport line. The 38 built were obsolete before the last was delivered in the fall of 1940. (USAF)

The Army Air Corps wanted an improved version of the Douglas B-18 and that is what North American gave them—the XB-21 "Dragon." Only one was built of the five ordered. Span was 95 feet. (USAF)

Pratt & Whitney Hornets of 1,200 horsepower each powered the Dragon. A .30 caliber machine gun was in the nose ball turret. In addition, there were two waist guns plus dorsal and ventral gun positions—all .30 caliber. (USAF)

Comparative three-quarter front views of the B-25 and B-26 show basic design differences between the two aircraft. (USAF)

The NA-40-2 (X-14221) was the precursor of the B-25. Only one was built. Powered by 1,350 horsepower Wright Cyclones, it had a maximum speed of 285 miles per hour. It was intended as a light attack bomber but the design was soon reworked into the B-25 Mitchell medium bomber. (North American)

The B-25 Mitchell and B-26 Marauder were, as were all mediums, a more personal means of attacking the enemy than the high-flying B-17's and B-24's. Strafing from tree-top level they bombed any likely target in sight and were in most cases low enough to receive a return barrage of small-arms fire directed at knocking them down. Even the lowliest infantryman with his service rifle had a pot shot at them.

Though the B-25 and B-26 medium bombers were built to the same Army Air Corps design requirements, if they were parked side by side it could make you wonder. Due to the perfectly round cross-section, the internal fuselage volume of the B-26 approached nearly twice that of the B-25 with its slender rectangular shape. The small, stubby wings of the B-26 looked almost racer-like compared to the more conventional wing-to-fuselage proportions of the B-25. The engine and nacelles of the B-26 were somewhat larger than those of the B-25. The main landing gear were even different—the B-25's folded *rearward* into the nacelles; the B-26's folded *forward* into the nacelles. The B-25's nose gear simply swung backwards while the B-26's turned 90 degrees before packing itself into its fuselage bottom. The landing-gear doors on the B-26 were open on the ground—on the B-25 they were closed. The B-26 propellers were four-bladed and the tips came close to the ground. On the B-25 they were three-bladed and missed the ground by a comfortable margin.

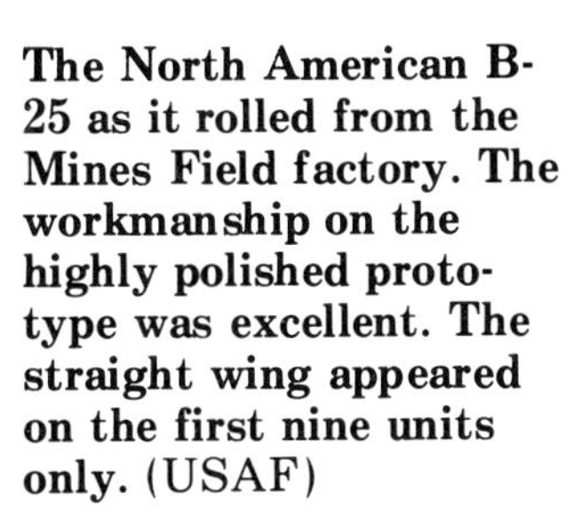

The North American B-25 as it rolled from the Mines Field factory. The workmanship on the highly polished prototype was excellent. The straight wing appeared on the first nine units only. (USAF)

Another of the initial batch of twenty-four B-25's—this one in olive drab with gulled wings to improve directional stability.

The B-25 design used many sheets of aluminum in its airframe and thousands of rivets to hold those sheets in place. The B-26 used much larger compound-curved thick aluminum plates which in some cases were welded in place instead of riveted. The B-26 had two bomb bays, each bay with different types of doors, while the B-25 had only one bomb bay. The B-25's many-paned bombardier's compartment contrasted to the smooth one-piece plastic enclosure on the B-26. Even the top turrets were different: the Bendix turret on the B-25 sat on a pedestal like a barber's chair, while the B-26 Martin turret hung from the top of the fuselage on a giant ball bearing. Lastly, the B-26, the faster of the two, was a real Gee Bee hot potato on landings and takeoffs while the B-25 was more staid and conservative, almost like a DC-3.

Back in the mid-1950's there were two primary schools of thought in U.S. car styling. General Motors products were rounded and soft on the edges and corners. Ford designers looked for a squared-off crispness in their offerings. So it was with the B-25 and B-26. Martin went for the rounded designs—the "Maryland" and "Baltimore" bomber designs just prior to the "Marauder" show this—while North American styled the B-25 crisply, sharp-lined and super functional. North American Aviation liked the "German" philosophy; the Ju 87 Stuka and Messerschmitt Bf-109 were examples of this approach and it certainly showed up in the P-51 Mustang of immortal fighter fame. The B-25 and B-26 *were* different, and not just in styling.

When the first B-25 rolled out of the North American factory at Inglewood's Mines Field it was a very charismatic, attractive and clean-limbed beauty. Compared to the NA-40-2 the fuselage was wider and longer, the span greater and it was fitted with larger engines. Yet its similarity to the earlier NA-40-2 was most apparent. On August 19, 1940, the prototype B-25 Mitchell's wheels felt air between them and the runway. Ninety-seven days later, on November 25, 1940, the first Martin B-26 Marauder took to the air.

Roosevelt had called on the American aircraft industry for 50,000 warplanes. Fortunately there existed a highly skilled nucleus of airplane manufacturers in this country that could be rapidly expanded into the sprawling complexes needed to produce so large a quantity of aircraft. Boeing, Martin, Douglas, Consolidated and North American were all highly accomplished and ready to commence the frantic pace of all-out war production.

North American Aviation had started as a large conglomerate of smaller aviation companies on December 6, 1928. It was a join-up of Wright Aeronautical, Curtiss-Robertson Airplane Manufacturing Company, Curtiss-Caproni Corporation, Curtiss Airplane and Motor Company, Moth Aircraft Corporation, Travel Air, and Keystone Aircraft Corporation—all going businesses. It was just after the romance of the Lindbergh epic, so capital was plentiful from Wall Street. Twenty-five million dollars were quickly raised with the sale of two million stock shares in the new venture. When the bubble burst in October 1929, this vast capitalization carried the new corporation through.

Although in styling North American Aviation could be compared to Ford, like General Motors, NAA had no pioneering founder's name such as Martin, Douglas or Boeing. Also like GM, it was run by a shrewd financial tactition who left the engineering and building to the employees. At GM the wizard was Alfred Sloan. At NAA it was Clement M. Keys, an ex-editor of the *Wall Street Journal.* Once formed, North American continued acquiring new additions, the more notable being Sperry Gyroscope, Pitcairn Aviation (the autogyro and mailplane people), Ford Instruments and Berliner-Joyce. It got into the airline business, owning big portions of TWA, Western Air Express and Eastern Air Transport for a time. The corporation also had other interests of a more minor nature.

James H. "Dutch" Kindelberger was appointed president of North American and he had a good background for the job. He had been an Army Air Service pilot toward the end of World War I and had worked as a designer and

This B-25A, one of 40 built, had increased armor plate for the pilots and gunners as well as self sealing fuel tanks in the wings. These aircraft were the first to enter operational service with the 17th Bomb Group.

The B-25A in a left bank shows off its large glassed-in prone tail gun position housing one .50-caliber flexible machine gun.

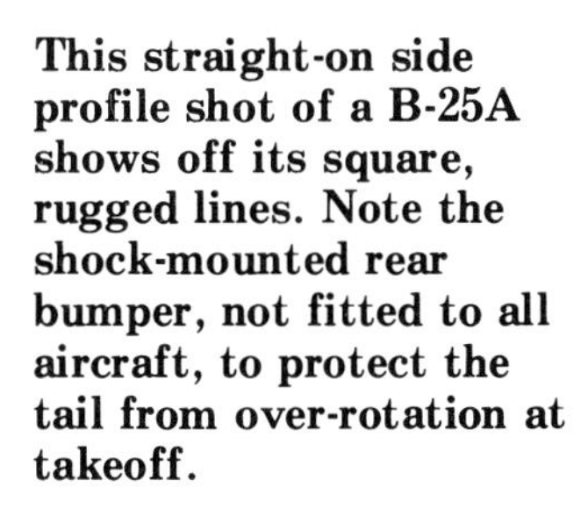

This straight-on side profile shot of a B-25A shows off its square, rugged lines. Note the shock-mounted rear bumper, not fitted to all aircraft, to protect the tail from over-rotation at takeoff.

The B-25A, without dorsal or ventral turrets, was a very clean airframe despite its lack of curves—a strictly functional, no frills design. (USAF)

A direct front view of a B-25A shows the zero dihedral of the outer wing panels. (USAF)

chief draftsman at the Martin plant until 1925. He then went on to become chief engineer of Donald Douglas's budding aeronautical company where he contributed to the development of the DC-2/DC-3 series. It all added up to the man at the top having a lot of savvy when it came to building great aircraft. He remained in this position until his death in 1962.

When he moved to North American, Kindelberger brought a trusted friend and excellent aero technician for the position of NAA chief engineer, John Leland Atwood, another Douglas alumni. (Not connected with Lee Miles's sensational *Miles and Atwood Special* of 1934 air racing fame.)

As war clouds gathered over the United States in the late 1930's, the War Department was interested in something along the lines of an improved Douglas B-18, and that's what North American gave them—the XB-21 "Dragon," with a few extras thrown in to make it unique. For altitude performance the two Pratt & Whitney 1,200-horsepower R-2180-1 Hornet engines were turbo-supercharged, and the ship had a complete oxygen system for the crew. It was quite heavily armed with five .50-caliber machine guns in separate positions including power-operated nose and dorsal turrets.

Work on the design started in 1935 and in March 1937 the prototype "Dragon" was rolled from the factory. In most respects it looked just like the B-18. Maximum speed turned out to be only 220 miles per hour; considering the Martin B-10 of 1934 could do 234 miles per hour, there wasn't much reason to get excited about the XB-21 in the speed category. Only one copy was built (serial number 38-485) and off it went to Wright Field for evaluation. The Air Corps' findings were quick and to the point: cancel the order for five additional YB-21's (after the first unit, the "X" designation was dropped). They took the nose and dorsal turrets off the prototype and let the project drop, using the plane only as a field hack.

It was probably just as well. The Douglas B-18 and B-18A series that did see limited production contributed little except training flight time for its crews.

The next great step for NAA engineering was the NA-40 aircraft from which the B-25 would eventually be developed. Even the negative experiences with the XB-21 paid off. The new design was crisp of line and looked good, and, as many designers will tell you, if it looks good it *is* good. It was built in response to USAAC Circular Proposal 38-385 which called for a new twin-engine light attack bomber. Chief Engineer Atwood and Raymond H. Rice headed up the project. Now they were finally digging where the gold was. A tricycle gear, shoulder wings and big, big engines for the size of the plane assured some "zappy" performance.

On roll-out in January 1939 at Inglewood, California, the highly polished, meticulously constructed ship was a real eye-popper. The graceful wing spanned sixty-six feet with an area of 598.5 square feet. The overall length was 48 feet 3 inches and the height 15 feet 2 inches. Two Pratt & Whitney R-1830 engines hung beneath the wings in neatly streamlined nacelles into which the

This view of the B-25 shows the flat dihedral of the outer wing panels and the boxy fuselage shape with extensive frontal greenhouse structure. (USAF)

This photo of a banking B-25B shows several changes that made it different than the A model, namely, the periscope-sighted Bendix ventral turret, the Bendix dorsal turret and the substitution of an observer's position in place of the tail gunner's station. Note the color line between the gray bottom and olive drab sides and top and streamlined housing over tail bumper. Of the 119 built, 24 went to Tokyo with Jimmy Doolittle. (USAF)

main landing gear also retracted. The engines were twin-row 14-cylinders, and each developed 1,100 horsepower. Twin tails placed the fins and rudders in the prop wash where it was figured they would do the most good.

A few flights by test pilot Paul Balfour showed the NA-40-1 to be as good in the air as it looked on the ground. To throw another 400 horsepower into the performance curves, the Pratt & Whitneys were pulled and 1,300-horsepower Wright GR-2600-A71 Cyclones were fitted. With this change the NA-40-1 became the NA-40-2.

By March of 1939 the prototype was at Wright Field for testing—and not without some great expectations. The ship now grossed out at 21,000 pounds and had a maximum speed of 285 miles per hour. It carried a crew of three and a bomb load of 1,200 pounds. Its competition in the test program was the slick Douglas A-20 (or DB-7), another hot-rod with similar design philosophies backing it up. Unfortunately, two weeks after its Wright Field arrival, pilot error destroyed the NA-40-2 prototype. Major Younger Pitts, on detached service from Barksdale Field Louisiana, was piloting the ship on an evaluation flight. He was not a regular test pilot and he lost control of the hot plane as he turned on final into Wright Field. The crew escaped but the NA-40-2 wreckage caught fire and was completely destroyed. But once more, it would not be a disadvantage.

The Army Air Corps brass thought the design was pretty good based on what they *had* seen. However, without a prototype they had to award the production contract to Douglas and the rival A-20 went into production. But the War Department, facing a global conflict, knew what its future bomber requirements were. Among them was the new medium bomber class. The Air Corps suggested that NAA take another look at their NA-40-2 design and uprate it to meet the medium bomber requirement. The result was the NA-62, brought about by Atwood and Rice at NAA engineering. The span was up a foot and a half with 12.5 square feet more wing area than the NA-40-2. Overall length was stretched to 52 feet 11 inches—almost five feet longer. The fuselage was broadened for side-by-side seating of the pilot and copilot and to accommodate 2,400 pounds of bombs—double what the NA-40-2 could carry. The cockpit canopy was now more closely integrated with the fuselage top. The wing was lowered from shoulder mount to mid-fuselage for better positioning of the mainspar in the fuselage structure. Wright Cyclone R-2600-9 engines of 1,700 horsepower each were fitted. Twin tails and tricycle landing gear were retained. On the drawing board it looked great—a quantum improvement over the already well thought-of NA-40-2.

The Air Corps couldn't wait—they wanted production of the design at once. On September 10, 1939, 184 NA-62's were ordered while only in blueprint form

The B-25D-NC (Mitchell II) was the fourth aircraft off the North American line at Kansas City. It was the same as the Mines Field built B-25C-NA model. These models featured a 24 volt electrical system and increased fuel capacity in the wings. (USAF)

This aircraft, a B-25D-30-NC, resides at the Air Force Museum. It is painted to resemble a Doolittle Raider complete with erroneous serial numbers. Note the short stub stacks around the cowling installed to hide the engine's exhaust plume at night. (Collect Air Photos)

A B-25G shows its major change from the C and D models—a shortened nose in which a 75 mm cannon reposes. (USAF)

for $11.7 million, or $64,000 each. By August 19, 1940, the B-25 Mitchell, named after airpower proponent General Billy Mitchell of 1920's fame, was ready for its first flight. It had required 195,000 man-hours of engineering and 8,500 drawings to get this far.

Flight tests were good and the static-test airframe delivered to the Wright Field torture chambers on July 4, 1939, showed up A-OK as it withstood the massive piles of sandbags trying to destroy it. Directional stability in the bombing run, while not atrocious, could have been improved. A new shape for the vertical fins and rudders was attempted but the aerodynamics people finally zeroed in on excessive wing dihedral as the culprit. The wing outer panels were rerigged to an almost dead-flat dihedral. That corrected the problem and the B-25 was off to the races—or, in this case, the wars.

Across the country in Baltimore, Maryland, another American aircraft firm was pegging away at the USAAC circular requirements for a medium bomber. The Glenn L. Martin Company's Model 179, which became the B-26 Marauder eventually, was under full design and development.

The Martin Company, like the Ford Motor Company, was named after its founder. The Aero Club of America had established a certificate for "expert" aviators in 1912 and Glenn L. Martin was awarded certificate number two. By 1914 only twenty-four such honors had been bestowed.

Glenn Luther Martin was born in Macksburg, Iowa, on January 17, 1886. Martin had achieved enough skill in building and flying by 1912 that he was often booked as the daredevil for county fairs and Fourth of July celebrations.

He was a real showman and publicity hound, getting writeups for chasing coyotes and escaped convicts from the air, opening "Pacific Aerial Delivery Route Number 1" for the U.S. Mail, dropping baseballs into catchers' mitts and flowers into beauty contest winners' hands from the air, and eventually playing the aviation hero opposite Mary Pickford in the 1915 film *The Girl of Yesterday.* Meanwhile, back at his small factory in an abandoned church in Santa Ana, California, he was still building a few bamboo-and-wire canvas-covered airplanes, complete with a flying school for prospective customers.

In 1916 the Wright-Martin company was formed through the merger of the original Wright Company, headed by Orville Wright (Wilbur had died of typhoid in May 1912); the Simplex Automobile Company, and the Glenn L. Martin Company. After the company disintegrated at the end of the war, Martin re-entered the airplane business with a contract dated January 17, 1918, for ten MB-1 twin-engine bombers. With two 400-horsepower Libertys hauling it along, the biplane could do close to 100 miles per hour.

Next came the MB-2, twenty copies of which were ordered by the Army in 1920. It was bi-winged and angularly ugly; and powered by two 420-horsepower Liberty engines mounted on the lower wing. The wing span was 74 feet 2 inches, the length was 42 feet 8 inches, with a height of 14 feet 8 inches. It weighed in at 12,119 pounds, had a service ceiling of 8,500 feet, range of 558 miles and blasted along at 99 miles an hour. These aircraft were the ones that General Billy Mitchell used to sink the supposedly impregnable captured German battleship *Ostfriesland* in just twenty-five minutes on July 21, 1921. (The story of General Mitchell's visionary fight for airpower, his court martial and his death in 1935, before the brilliance of his plans could come true, is indeed sad. An ironic aspect of all this was that he had made his name in aerial bombardment flying Martin aircraft, yet the plane that carried his name through World War II was the head-on competitor of the Martin B-26, the North American B-25 Mitchell.)

The name Martin became big on the American aeronautical scene, with a roster of designs for both the Army and Navy being produced one after the other during the 1920's and early 1930's. There were the T3M, T4M and TG series of biplane torpedo bombers built for the U.S. Navy, comprising, for the 1920's, good-size contracts. The T3M series numbered 124 aircraft delivered in 1927 and more followed, with 102 T4M's, and fifty TG-1 and TG-2's. T3M-2's were the Navy's first torpedo bomber squadron aboard the carrier *USS Lexington* in December 1927. In the late summer of 1929 the first P3M flying boat hit the water. With a 100-foot span and twin-engined, it was capable of making the hop to Hawaii from California, nonstop, 2,350 miles. While the design was by Consolidated (as was the prototype), Martin underbid them and made off with a contract for nine P3M's. Three Martin four-engined giants were built and put into service—the Model 130. First was the very famous *China Clipper,* followed by the *Philippine Clipper* and the *Hawaii Clipper.* They wound up, after four years on the Pan-American Pacific route network, in the U.S. Navy during World War II as long overwater transports. The span was 130 feet and length was 90 feet 7½ inches.

When the era of the full-cantilever monoplane arrived, Boeing was pushing its Monomail and its YB-9 twin-engine bomber with full-retract gear. The large bomber prototype could fly faster than the first-line P-26 "Peashooter" fighter. Martin jumped on this one with a vengeance and came up with its company-funded Model 123 in 1932. With an all-metal full-cantilever span of 70 feet 10 inches, twin 675-horsepower R-1820-F Cyclones, a rotating-nose turret, a retract-gear and a fighter-like speed of 207 miles per hour, the Army bought the design on January 17, 1933. With improvements it stayed in production through 1936, and like "Piper Cub" meant *any* high-winged light plane to the uninformed, "Martin bomber" meant *any* bomber during the mid thirties. Production included the XB-10, one unit; the YB-10, fourteen units; the YB-10A, one unit; the B-10B, 103 units; the YB-12 (B-10 with larger engines), seven units; the B-12A, twenty-five units; and the XB-14 with Twin Wasp engines, one unit. In addition to this formidable production, another 206 samples were built for export up to 1939: Russia, one; Siam, twenty-three; China, nine; Turkey, twenty; Argentina, thirty-five; and a whopping 118 for the Netherlands East Indies, which operated them against the Japanese early in World War II though they were by then totally obsolete.

While all this was going on, Martin continued to court the Navy with its PBM flying boat, and PR-wise they had the inside track. *National Geographic* magazine had covered the *China Clipper* saga in great and complimentary detail—and who can turn down a real winner!

Twenty production PBM-1's were ordered into production in December 1937 for 1940-41 delivery. The big boat, with twin 1,600-horsepower Pratt & Whitneys, highly dihedraled twin-tails and a massive pregnant hull, had a span of 118 feet and a length (depending on the dash number) of 77 feet 2 inches to 79 feet 10 inches. The production of this basic design was to continue right up to V-J Day, when 460 were canceled due to the end of the war. A last variant of this job did survive, however—the PBM-5A with big, big R-2800-34 engines and a retract-gear, making it "super amphibian." They were delivered in 1949.

Martin was on good terms with the Army Air Corps. Glenn Martin, still at the helm, had many friends at the top of the Army Air Corps brass stack: Arnold, Doolittle, et al. One Martin design caught the eye of the now-beleaguered British and French—the Model 167. It was Martin's reply to an Army circular for a new twin-engine attack aircraft. (However, the Douglas A-20 [DB-7] won.) France ordered 115 of them. It was a twin-engined hot-rod light bomber. When France fell to the Germans the little bombers were diverted to Britain and became the "Maryland I" and more powerful "Maryland II." They served the British in the Middle East during 1940-41 as recon and light bomber aircraft. They were slender, mid-winged, twin-engine types with 1,200-takeoff-horsepower Wasp radials. They were lightly armed with .30 calibers, four in the wings, a ventral and a dorsal gun, and as such never really made much of a splash in World War II. The Maryland had a wing span of 61 feet 4 inches and length of 46 feet 8 inches with the two Pratt & Whitney R-1930's (1,050 horsepower each) dragging it along. Normal bomb load was 1,250 pounds.

It can be readily seen that Glenn L. Martin Company—under the close watch of Glenn L. Martin—was a viable and hard-hitting member of the U.S. military aircraft industry. They were used to multi-engined bomber aircraft, and all-around quick reaction time to the military's needs. So it was in the case of the B-26.

The B-26 story started with the Material Division at Wright Field issuing Circular Proposal 39-640 on March 11, 1939. It went to over forty manufacturers—anyone the Army thought would even be remotely interested in building a medium bomber—a new class falling between the Douglas A-20 and the B-17. Though range and service ceiling could be less than the Fortress, they wanted about the same bomb load and a 300 flat-out miles-per-hour top speed. The last item was tough, for even first-line fighter planes were not going much faster than that when laden with a full combat load.

Glenn L. Martin, always on the lookout for new business, snatched up the challenge. While not an engineer himself, he ran the show at the airplane works. He would continue with his firm hand in the business until his death of a cerebral hemorrhage on December 4, 1955, near age seventy. He was backed up by a crack team of engineering talent including William E. "Ken" Ebel, chief engineer and superb test pilot as well. Next came Executive Engineer W. T. Willy, later to take over the Omaha, Nebraska, plant as vice president and general manager. To head up the actual work on the Martin proposal in answer to the circular, a project engineer was named: Peyton Marshall Magruder, a twenty-eight-year-old graduate (1934) of the U.S. Naval Academy. After leaving Annapolis he had taken an aeronautical engineering degree from the University of Alabama, then went to work at the Naval Aircraft Factory. That led

ENG. REP. NO. 1099

GLENN L. MARTIN COMPANY

MODEL 179

SUMMARY OF PERFORMANCE

AIR CORPS CIRCULAR PROPOSAL 39-640

BID NO.	UNITS	1	2	3	4	5	6	7	8	9	10	11	12	13	14	15
ENGINE		WAC - 2600					P & W - 2800						WAC - 3350			
ENGINE TYPE		1 STAGE 2 SPEED	1 STAGE 2 SPEED	1 STAGE 2 SPEED	2 STAGE 2 SPEED	TURBO	1 STAGE 2 SPEED	2 STAGE 2 SPEED	TURBO	1 STAGE 2 SPEED	2 STAGE 2 SPEED	TURBO	1 STAGE 2 SPEED	2 STAGE 2 SPEED	TURBO	TURBO
ENGINE NOSE (SHORT, LONG)		SHORT	LONG	LONG	LONG	SHORT	SHORT	SHORT	SHORT	LONG	SHORT	SHORT	SHORT	SHORT	SHORT	SHORT
ENGINE SPEC. NO. AND DATE		B-655-3/6-6-39	B-655-1/6-6-39	B-655-1/6-6-39	B-658-1/6-10-39	B657-4/6-10-39	A-8019/5-15-39	A-8020/5-15-39	A-8021/5-15-39	A-8024/5-15-39	A-8025/5-15-39	A-8026/5-15-39	B-665/6-17-39	B-669/6-17-39	B-670-1/6-17-39	B-670-1/6-17-39
TAKE OFF HP/RPM		1700/2600	1700/2600	1700/2600	1700/2600	1700/2600	1850/2600	1850/2600	1850/2600	2000/2700	2000/2700	2000/2700	2200/2600	2200/2600	2200/2600	2200/2600
NORMAL RATING - ENG. CRIT. ALT. / HP/RPM		13000/1350/2400	13000/1350/2400	13000/1350/2400	22000/1290/2400	25000/1500/2400	13000/1450/2400	21500/1460/2400	25000/1500/2400	13000/1450/2400	21500/1550/2550	25000/1625/2550	15500/1800/2400	18200/1725/2400	25000/2000/2400	25000/2000/2400
MILITARY RATING - ENG. CRIT. ALT. / HP/RPM		12000/1450/2600	12000/1450/2600	12000/1450/2600	21500/1388/2600	25000/1700/2600	14000/1500/2600	21000/1600/2600	25000/1850/2600	13500/1600/2700	19500/1800/2700	25000/2000/2700	16000/1900/2600	20000/1850/2600	25000/2200/2600	25000/2200/2600
ENGINE WEIGHT / PROPELLER WEIGHT		2000/594	2105/594	2105/476	2210/476	1985/476	2265/594	2375/476	2210/476	2505/476	2496/476	2320/476	2500/604	2630/604	2480/604	2480/604
PROPELLER GEAR RATIO		2:1	2:1	2:1	2:1	2:1	2:1	2:1	2:1	2:1	2:1	2:1	16:7	16:7	16:7	16:7
PROPELLER DIA. / NO BLADES / σ (.7 RADIUS)		13.5/4/.127	13.5/4/.127	12.5/4/.117	12.5/4/.117	12.5/4/.117	13.5/4/.127	12.5/4/.117	12.5/4/.117	12.5/4/.117	12.5/4/.117	12.5/4/.117	13.5/4/.127	13.5/4/.127	13.5/4/.127	13.5/4/.127
PROPELLER DESIGNATION		C-543S - 814Cc2	C-543S - 814Cc2	C-542S - 714Cc2	C-542S - 714Cc2	C-542S - 714Cc2	C-543S - 814Cc2	C-542S - 714Cc2	C-542S - 714Cc2	C-542S - 714Cc2	C-542S - 714Cc2	C-542S - 714Cc2	C-643S - 814Cc3	C-643S - 814Cc3	C-643S - 814Cc3	C-643S - 814Cc3
PROPELLER SPEC. NO.		P-39	P-39	P-36	P-36	P-36	P-39	P-36	P-36	P-36	P-36	P-36	P-41	P-41	P-41	P-41
WEIGHTS																
NORMAL GROSS WEIGHT	LBS.	24700	25581	24840	25140	25630	26087	26140	26140	25830	26538	26540	27261	27725	27725	28232
WEIGHT EMPTY	LBS.	18174	18464	18132	18767	18784	18958	19285	19421	19164	19366	19480	19664	20231	20554	20635
USEFUL LOAD	LBS.	6526	7117	6708	6373	6846	7129	6855	6719	6666	7172	7060	7597	7494	7171	7597
FUEL / OIL	LBS.	2250/254	2780/315	2412/274	2110/241	2535/289	2790/317	2546/287	2422/275	2374/270	2829/323	2727/311	3210/365	3118/354	2827/322	3210/365
USEFUL LOAD LESS FUEL AND OIL	LBS.	4022	4022	4022	4022	4022	4022	4022	4022	4022	4022	4022	4022	4022	4022	4022
PERFORMANCE																
1. HIGH SPEED AT NORMAL RATED CRIT. ALT. + 2000 FT. RAM	M.P.H.	315	313	320	335	355	322	349	354	327	355	366	360	365	402	401
$HP/N_N/V_T/V_C$		1350/.824/.915	1350/.824/.911	1350/.858/.866	1290/.822/.892	1500/.792/.923	1450/.824/.915	1460/.811/.914	1500/.793/.923	1450/.859/.868	1550/.788/.957	1625/.769/.970	1800/.845/.869	1725/.822/.883	2000/.798/.930	2000/.798/.930
2. HIGH SPEED AT MILITARY RATED CRIT. ALT. + 2000 FT. RAM	M.P.H.	317	315	323	338	368	324	355	380	335	368	392	365	377	414	413
$HP/N_N/V_T/V_C$		1450/.794/.970	1450/.795/.970	1450/.835/.918	1388/.788/.954	1700/.760/.990	1500/.793/.980	1600/.783/.979	1850/.764/1.000	1600/.822/.958	1800/.769/1.000	2000/.743/1.032	1900/.815/.926	1850/.794/.942	2200/.785/.990	2200/.785/.990
3. HIGH SPEED AT 15000 FT.	M.P.H.	315	313	320	314	324	322	329	324	327	336	334	352	345	365	364
$HP/N_N/V_T/V_C$		1350/.824/.915	1350/.824/.911	1350/.858/.866	1350/.819/.859	1500/.807/.868	1450/.824/.915	1540/.821/.868	1500/.807/.864	1450/.859/.867	1625/.803/.871	1625/.790/.870	1800/.845/.846	1685/.839/.847	2000/.810/.866	2000/.810/.866
4. SPEED AT 75% HP	M.P.H.	295	293	298	317	314	292	323	312	296	327	325	338	350	360	359
$HP/N_N/ALT.$		1125/.846/15000	1125/.846/15000	1125/.872/15000	1125/.827/24000	1125/.808/25000	1125/.846/15000	1200/.830/23500	1125/.807/25000	1125/.872/15000	1250/.820/23500	1220/.801/25000	1500/.869/17500	1500/.862/20200	1500/.826/25000	1500/.826/25000
5. OPERATING SPEED AT 15000 FT.	M.P.H.	227	275	254	250	244	269	255	252	260	250	260	300	295	263	292
$HP/RPM/N_N$		650/1550/.868	960/2090/.850	780/1880/.878	800/1900/.844	800/1900/.825	925/2080/.854	850/1750/.838	850/1750/.820	850/1950/.878	830/1750/.838	910/1850/.819	1190/2050/.850	1150/2000/.850	950/1725/.824	1170/2020/.823
6. RANGE AT OPERATING SPEED - 1/2 FUEL OVERLOAD	MI. ST.	1800	1800	1800	1520	1768	1800	1670	1660	1620	1800	1800	1800	1800	1800	1800
7. FUEL FOR 3000 ST. MILE RANGE AT ECONOMICAL SPEED	LBS.	7200	7310	7370	7825	8000	7690	8780	8875	7775	8990	8460	8560	8870	9270	9295
8. SERVICE CEILING	FT.	26500	26000	26400	31100	34700	27100	31900	33900	27200	31800	33600	30200	31000	36800	36300
9. SERVICE CEILING (SINGLE ENGINE)	FT.	13900	12600	12000	12000	13750	13840	14000	12500	12000	15000	14200	16500	14400	17800	17080
10. TAKE OFF & LANDING OVER 50 FT. OBSTACLE	FT.	2000	2200	2480	2360	2500	2100	2500	2500	2320	2500	2500	2390	2500	2500	2630
TAKE OFF DISTANCE / LANDING DISTANCE	FT.	2000/1980	2200/2050	2480/1990	2360/1900	2500/1930	2100/2080	2500/1970	2500/1970	2320/2070	2500/2120	2500/2120	2390/2170	2500/2200	2500/2200	2630/2240
CHARACTERISTICS																
WING AREA / ASPECT RATIO	SQ.FT./–	600/7.05	600/7.05	600/7.05	650/7.05	650/7.05	600/705	650/7.05	650/7.05	600/7.05	600/7.05	600/7.05	600/7.05	600/7.05	600/7.05	600/7.05
TAIL AREA / EXPOSED WING AREA	SQ.FT.	200/503.5	200/503.5	200/503.5	200/550	200/550	200/503.5	200/550	200/550	200/503.5	200/503.5	200/503.5	200/503.5	200/503.5	200/503.5	200/503.5
FUSELAGE AREA / ROOT CHORD	SQ.FT./FT.	46.1/14	46.1/14	46.1/14	46.1/14.5	46.1/14.5	46.1/14	46.1/14.5	46.1/14.5	46.1/14	46.1/14	46.1/14	46.1/14	46.1/14	46.1/14	46.1/14
NACELLE AREA / AV. WING THICKNESS RATIO	SQ.FT./%	35.2/15	35.2/15	35.2/15	35.2/15	35.2/15	31.7/15	31.7/15	31.7/15	31.7/15	31.7/15	31.7/15	35.2/15	35.2/15	35.2/15	35.2/15
FULL SCALE C_{D_0}/L		.0231/.0597	.0231/.0597	.0231/.0597	.0221/.0597	.0221/.0597	.0231/.0597	.0221/.0597	.0221/.0597	.0231/.0597	.0231/.0597	.0231/.0597	.0231/.0597	.0231/.0597	.0231/.0597	.0231/.0597
PERCENT COOLING HP - HIGH SPEED / CLIMB	%	0/4	0/4	0/4	4/6	6/8	0/4	4/6	6/8	0/4	4/6	6/8	3/6	4/6	6/8	6/8

NOTES:
1. MAXIMUM BUILT IN FUEL CAPACITY FOR ALL AIRPLANES - 1070 GALS.
2. BOMB BAY FUEL TANK CAPACITY FOR 3000 MILE RANGE AT ECONOMICAL SPEED AS REQUIRED.
3. ALL AIRPLANES SET UP WITH 4-600 LB BOMBS (2507 LBS) AND 5 MAN CREW.

Page one of the Martin bid for its Model 179 in answer to the Air Corps Circular Proposal 39-640 for a new medium bomber. Dated July 5, 1939, it detailed 15 variations on the theme with speeds ranging from 315 to 413 miles per hour dependent on the engines used. (USAF)

One of the eight general arrangement three-views included in the Martin bid. All had the twin tail which was in vogue at the time, e.g. the B-24 and B-25. (USAF)

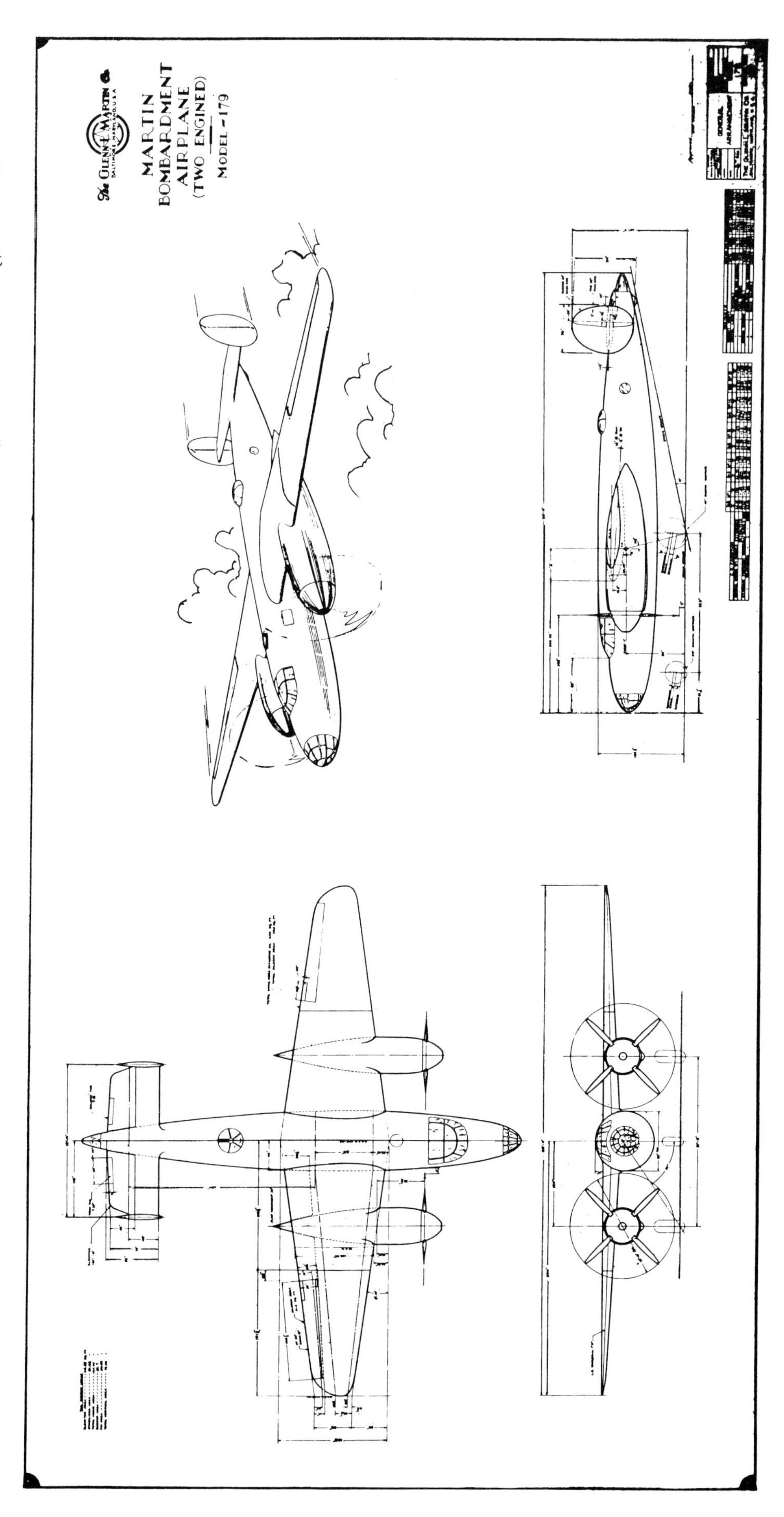

to several years of tutelage under top-drawer aircraft designer William H. Miller. By the time he arrived at Martin he was a pretty imposing package, education *and* experience-wise. Under Magruder's coordination other crack engineers worked on the Model 179 proposal. These specialists included Ivan Driggs, Carl Hartgard, Homer Huey, Fred Jewett, Clifford Leisey, James Muny, Clifford Roberts and George Trimble.

Right off the bat there was some disagreement amongst the team as to just exactly how to interpret the Air Corps circular requirements. Designing an airplane is always a compromise. To get off on the right foot, Chief Engineer Ebel went to Wright Field for clarification. He got it. They wanted high speed and they wanted bomb load, both items pushed for by Charles Lindbergh after he viewed the prewar Luftwaffe capabilities. Magruder pondered the proposal and realized the Wright Field people hadn't said anything about maximum landing speed. With design direction now set, Magruder and team went to work and a tank-car-load of midnight oil was burned before Martin submitted its bid on July 5, 1939. Bound in a navy-blue cover with the Martin logo in full color, the bid was actually a sixty-five-page book covering fifteen alternate design possibilities.

Basically the Model 179 proposal was a torpedo-bodied, twin-tailed, twin-engined, tricycle-geared machine with the variations centering around what speed range the Air Corps might want. The bid began with a three-page index, and a quick outline of it can provide some notion of what it contained. Page one was a chart showing the fifteen variations on the design and a summary of their performance. Following this were eight general arrangement drawings and display model photographs. As some of the major differences from model to model were internal, eight three-views covered the fifteen models externally.

The next section consisted of five drawings covering the inboard profile and fixed equipment, followed by a section on wing structure. Fuselage structure drawings were next, then a section of landing gear structural drawings and controls. The next two sections were structural details and flight control illustrations followed by a nine-page section on various engine, fuel and oil system installations. Near the end were various armament, bomb-bay and turret specifications. The bid ended with eight pages of options covering such things as alternate bomb-bay arrangements, armament, making self sealing fuel tanks for turbo and non-turbo bids, clamshell tail guns, emergency flotation bags, and a retractable upper rear turret. It was a Sears, Roebuck bomber catalog for the Air Corps.

The Martin Model 179 was top dog by far and the Air Corps opted for 201 of the machines for $16 million—$79,602 apiece. That was about $16,000 more per copy than the NA-62, which also got an order for 184 aircraft. No one else in the contest got the time of day.

The project was sold; now for a prototype and production. So far the concept was paper drawings and specifications. The first thing to change was the "in vogue" twin tail. The B-24 had it, the B-25 had it, the Navy's PBM had it and, of course, the Germans, Japanese and British were using it. It was a fad, to the mind of hard-headed Chief Engineer Ken Ebel, and it presented *two* vertical tails to miss blowing off, instead of one, with the dorsal power turret. Magruder agreed and the two of them personally lofted a new single vertical tail surface. By October 7, 1939, the Army had agreed and approved.

They soon found their sophisticated Model 179 (now B-26) was not quick and easy to build. It took 38,900 man-hours to roll one out the factory door versus, say, only 33,000 for a much bigger B-17. Parts and equipment vendors were another problem. After much effort, Magruder had Rohm and Haas producing the giant plastic nose, Aluminum Company of America forging a connection piece for the fuselage/wing junction larger than had been attempted before, and raised hell with General Electric trying to get them to move up the production schedule for an experimental electric motor for the gun turrets. By

A Martin artist's rendering of the new B-26 after the design was switched to a single vertical tail. Note that nose at this time was still a multipaned greenhouse. (USAF)

The first B-26 sitting on the ramp looking very clean and capable with its butt-jointed polished skin. The big Pratt & Whitney 18-cylinder radials made up a large part of the bomber's frontal area. (USAF)

Number one B-26 during taxi tests with Martin Chief Engineer Ken Ebel at the controls. It appears even nicer looking than the previous artist's drawing had anticipated. (USAF)

The original B-26 turning right with nose wheel and rudder during taxi tests. (USAF)

November 25, 1940, however, the super-sleek B-26 had been rolled out and taxi tested. All seemed go.

Chief Engineer Ken Ebel was pilot, Bob Fenimore was copilot and Al Melewski flight engineer. The prototype lifted off for a successful first flight November 23, 1940. The only thing found wrong was minor and easily corrected—there was a slight aerodynamic overbalance in the rudder. The prototype, after many more flight tests, did the unbelievable. It was pressed into service as a flight trainer by the USAAC!

The very talented Magruder went on to become a novelist and playwright in addition to being a pilot, aerodesigner and industrialist. As far as the B-26 program was concerned he considered his greatest achievement to be the design, building, and testing of the dorsal power turret in a mere three weeks—along with the help of Homer Huey. Though Glenn L. Martin initially took a dim view of attempting the job in so short a time, it soon became a best-selling Martin product—56,000 turrets were built and installed on twenty-seven different types of aircraft. The B-26 was now designed, flown and deemed a success. Like the B-25, it was off to the wars and trial by fire for this aircraft as well.

Top speed ranged from 315 to a wild 414 miles an hour—pretty much dependent on the engine horsepower and turbocharging. Three basic engines had been considered, the Wright R-2500, the Pratt & Whitney R-2800 and the Wright R-3350. The Wright R-2500 (1,700 horsepower) was well established. It was used on the North American B-25 but the Martin team felt it was too small—horsepower equals speed. However, the first five of these proposals used this engine with top speeds between 315 and 368 miles per hour. The second engine, the Pratt & Whitney R-2800, ranged from 1,850 to a 2,000 takeoff horsepower rating dependent on if it was turbocharged or not. That engine, with its variants, covered six additional Model 179 versions having top-speed ranges of 324 to 392 miles per hour. The wilder bids included four Model 179 designs using Wright R-3350-18 Duplex Cyclone engines that could belt out 2,200 horsepower—as much as a diesel locomotive! The problem was the engine was still under development even though Martin was hoping to use it on its giant 200-foot-span *Mars* flying boat. Speeds for the Model 179 with the behemoth powerplant ranged from 365 to a whopping 414 miles per hour, better than a P-38! The Army was aware of the engine pros and cons, too. They prudently elected for the Model 179-11 with Pratt & Whitney R-2800 turbo engines of 2,000 horsepower which provided a top speed of 392 miles per hour. That certainly should be good enough for a bomber!

Other items that kept Magruder and team busy were lesser, though important, requirements in the original proposal circular. For instance, the Army wanted the pilot to be able to see both wing tips from the cockpit for formation flying and general handling reasons. That dictated where the cockpit would have to be located in relation to the rest of the aircraft. With the high airspeed required, a racer-like small wing was needed to cut drag. It made the visibility requirement more difficult as the wing tips were not that far out from the pilot's seat and, therefore, were more difficult to see.

The low-drag symmetrical airfoil section seemed to be ideal streamlining, so everything else became that shape also—the fuselage, nacelles and tail mem-

The third B-26 with red cowlings went to the 22nd Bomb Group along with 57 others and they were soon on their way to Australia. (USAF)

An early B-26, now with dorsal turret, cruises along over the Maryland countryside on a test flight. (USAF)

A low-level pass by the prototype exudes speed, beauty and a quantum jump in the field of high-performance aerodynamics. (USAF)

bers. Due to the small wing (smaller than the lighter B-25) landing speed was way up, pushing over 100 miles an hour compared to the 60 to 70 miles per hour of other contemporary bombers. That dictated the tricycle gear; at least it wouldn't nose over with hard braking.

On July 17, 1939, twelve days after Martin's design proposal submission, the Air Corps Board of Officers met and picked the winners of the medium bomber competition. Martin (and Magruder) came out head and shoulders above everyone else—even North American, who had been encouraged by the Air Corps to hop up its NA-40-2 design into the new medium-bomber category. A point scale had been used by the board and here is how the major contestants came out by figure of merit:

	Points
Martin Model 179	813.6
North American NA-62	673.6
Douglas B-23	610.3
Stearman Model D-23	442.7

The Marauders of the 386th Bomb Group line up for inspection along with a Lockheed AT-18. (USAF)

This 1943 photo of a B-26C shows an aircraft fit and ready for action. Location was Meeks Field. (USAF)

3

ANATOMY OF A MITCHELL

WHILE ITS LINES WERE a bit on the square side, the North American B-25 Mitchell had a certain slender grace that made it a favorite with pilots, air crews and ground crews alike. Visibility was excellent and the gull-winged aircraft had no vices. This made for a stable bombing and gunnery platform and at the same time allowed a low-time pilot to make a mistake and still get out of it with his neck. The design was uncomplicated with regard to equipment and structure, making for easy maintenance and fairly easy modifications. The B-25 was the most widely used bomber of World War II. It was flown in all combat theaters by the Allies. It was a very fine aircraft and some detail concerning its specifications, performance, structure and equipment is in order. The B-25J will be generally used for the description, as it represents the aircraft developed to its highest combat state. That model also was produced in the largest quantity of any of the variants, 4,318 in all. They were built at North American's Kansas City plant between December 1943 and August 1945. A few of them went to the Navy as the PBJ-1J.

The B-25J model had evolved into almost a giant fighter bomber—its mission was to bomb and strafe land and naval material targets in support of ground, air or naval forces. However, a few restrictions (no loops, rolls, spins, inverted flight, vertical banks or Immelmans) kept it firmly classed as a medium bomber. Russian pilots of the B-25 paid little attention to these restrictions and happily popped rivets all over the place as they careened around the skies. It carried a six-man crew—up one from previous models—including pilot, copilot, bombardier and three gunners. The copilot did double duty as navigator while one of the waist gunners served the radio-operator slot.

The fuselage was of great importance in the aircraft for it carried the men, bombs, armament and equipment. Its skeleton was built up of 64 full and partial formers, which gave the fuselage its cross-sectional shape. They were spaced and riveted together with 36 aluminum Z-stringers. There were 10 stringers on the fuselage top, 9 on each side and 8 along the bottom. These, plus the formers, gave the fuselage its proper side-view profile. To help stiffen this relatively flimsy framework there was an aluminum I-beam at each fuselage bottom corner and a channel beam at each top corner. In addition, the crew floor helped to make the assembly more rigid. However, the real structural strength came when the aluminum sheeting was riveted to the skeleton, forming a very strong semi-monocoque assembly. Maximum height of the fuselage was 7 feet 4 inches and the maximum width was 4 feet 8.5 inches. Overall fuselage length was 52 feet 11 inches. This structure was comprised of 23 major subassemblies (some made by companies other than NAA) which were mated to form the whole on the final production line.

Anyone wondering why occupant enclosures on aircraft are sometimes called "greenhouses" has only to look at the bombardier's enclosure on the B-25. The nose, made up of 23 transparent panes mounted in an aluminum framework, gave the bombardier an excellent view down and ahead for bomb-aiming

Three-quarter rear views of B-25J-NA show the new tail gunners' turret and bay window waist gunners' positions adapted from the B-25H model. (USAF)

The B-25J-NC (Mitchell III) was the last and most produced variant of the B-25's—4,390 J's in all. This photo shows package guns on fuselage side and forward mounted dorsal turret. (USAF)

A fine view showing all the gun positions of a B-25J-NC—twelve .50-caliber machine guns in all. (USAF)

and a wide overall view for manning the hand-held .50-caliber Browning to ward off frontal attackers. Two other greenhouses, the pilot and copilot's canopy assembly with 15 panes and the raised tail-gunner's position with seven panes, completed the high-visibility crew stations. A first ride in the tail gunner's position was always a little unnerving for the occupant. As viewed from that spot, the tail assembly seemed to virtually spread from the gunner's shoulders and gave the viewer a real closeup of the aerodynamic flexing of the tail structure in rough air. More than one uninitiated rider in this position was sure the tail was going to break off at any second.

Two large waist-gun windows, almost bay windows due to their shape, were located on the fuselage sides behind the wings. The one on the right side was well ahead of the left one when viewed from above the aircraft. This gave the gunners more room in the narrow fuselage to swing their respective .50's through firing arcs without interfering with each other. Additional openings into the ship were the front and rear crew entrance hatches. The forward hatch, located on the fuselage bottom, entered into the navigator's compartment. The rear hatch, also on the bottom, was located just behind the waist-gunner's station. Steps were built into the hatch covers. For an emergency exit, a third manhole-type hatch located in the crawlway floor over the bomb bay was available. Other openings, closed off by doors during flight, were the nose-gear doors which were normally closed on the ground, but opened long enough for the nose wheel to go into its well during retraction before snapping shut behind it. This was accomplished with a mechanical connection between the nose-gear strut and the doors. Another set of doors covered the bomb bay and were operated hydraulically by controls in the bombardier's compartment. They were never to be opened at speeds over 290 miles per hour.

The wing structure was 15 premanufactured subassemblies including the massive center section, two outer panels, two tips, four flaps, two ailerons and four leading edge sections. Fully assembled, the wing was of two-spar construction with a third stub spar reaching out as far as the nacelles to help carry the main landing gear loads. The airfoil shape was made up of 28 ribs in each wing, starting with a NACA 23017 section at the root and blending outward to a 4490R section at the tip. This resulted in a thick high-lift center section with a more symmetrical streamline section at the tip. Fifteen stringers were used in the center section, with 11 on the outboard panels—the whole assembly then being covered with stressed sheet aluminum. The wing was mounted to the fuselage with a 3° 30' angle of incidence. The gull effect was achieved with the center section dihedral at 4° 38' 23" and the outboard panels set at negative 0° 21' 39". The wing chord at the fuselage was 12 feet 10.8 inches, which, with the 67 ft. 6.704 inch span, gave an aspect ratio of 7.48.

The ailerons were all-metal frames with doped fabric covering, mounted to the wings with three hinges. The aileron trim tabs were sheet aluminum. Aileron travel was 28 degrees up and 14 degrees down, controlled by duplicate control cable systems. The dual system was also used with elevators and rudders to provide control of the aircraft even if one cable system was damaged in combat. The trim tab travel was plus or minus 12 degrees.

The flaps consisted of an inboard and outboard section on either side of each engine nacelle and they could be set at any angle from closed to down 45 degrees. They were hydraulically controlled by means of a lever on the pilot's control pedestal. They were generally extended 20 degrees for takeoff and full for landing, but were not to be extended at air speeds above 170 miles per hour. In the event they would not lower hydraulically due to battle damage or mechanical malfunction, the pilot could ask the waist gunner/radio operator to lower them by means of an emergency crank in that compartment. The operation took 27 turns of the crank so time during the approach had to be allowed for the manual operation. The flap area was 75.8 square feet of the total wing area of 609.8 square feet. Nearly 40 inspection plates on the wing undersurface allowed access to all major mechanisms within the wing. The right wing tip supported a pitot tube which extended forward approximately two feet.

The twin tail assembly spanned 22 feet 8.5 inches with a maximum chord of 6 feet 10 inches and was set at two degrees incidence, providing a slightly nose-down moment. Area, including elevators, was 132.4 square feet, the elevators making up 40.4 square feet. Elevator travel was 25 degrees up and 10 degrees down. The reason for the greater up travel was its use in flaring-out for a main-wheel landing at low air speeds. Structure of the stabilizer was two-spar with 18 ribs total. It was covered with sheet aluminum in semi-monocoque fashion. The elevators were hinged on either side of the fuselage with two hinges. They were metal framed, using 13 ribs in each panel, and were fabric covered. Each elevator had a 2.11-square-foot trim tab, all-metal, with a four-degree up travel and 20-degree down travel.

The rudder and fin assembly had a height of 102.75 inches; each fin had an area of 57.4 square feet and each rudder had 33.6 square feet. The rudders were metal framed, fabric covered, and were attached to the all-metal fins at four hinge points each. Rudder travel was 20 degrees in either direction; trim tabs were 12 degrees in either direction.

The final components of the airframe were the engine nacelles. Including the cowl, they were made of seven major subassemblies. Fourteen formers gave them their cross sectional shape and, like the fuselage, the use of Z-stringers provided the side and top profiles. Sheet aluminum riveted to this framework produced a neatly streamlined enclosure to house the retracted main landing gear. The two main gear doors were mechanically connected to the landing gear struts and, as with the nose wheel doors, remained closed at all times except during actual retraction. Having the doors closed during takeoff reduced drag, and prevented damage from stones and mud to the inside of the nacelles. Length from the cowling nose to the rear tip of each nacelle was about 19 feet.

It should be noted that to bring together from various points of the country nearly 50 different major subassemblies and have them go together to form a complete major aircraft, ready for combat, required extraordinary engineering and precision building. On the original B-25 prototype, 195,000 engineering man-hours and 8,500 drawings were required. To get to the same point on the J model easily required twice that.

An airframe alone is nothing but a sturdy, polished metal sculpture until it is given some muscle—the engines. In the case of the B-25J, there were two Wright Cyclone R-2600-13 (or R-2600-29) 14-cylinder mechanical marvels that gulped 100—130 octane aviation gasoline and produced 1,700 takeoff horsepower at 2600 rpm in return. When throttled back for high-speed cruise at 2400 rpm they still produced 1,500 horsepower each for as many hours as the fuel supply lasted. At takeoff horsepower, the pilot was limited to five minutes of flat-out operation before cylinder head temperatures reached 500°F, after which lubrication failed and physical damage was done to the engines. Head temperature at high-speed cruise ran about 428°F. If long range was desired, revolutions per minute were dropped to 2100, bringing the head temperature down to 396°F. (Aircraft engines, designed for lightness and extreme horsepower, require a great deal more operating attention than do automobile engines. Automobile engines do not run at maximum horsepower for any significant period, if at all, and weight is not a major consideration. They therefore can be built structurally more beefy to withstand relatively low stress levels and provide reliability.)

Since the Wright Cyclones were air-cooled, cowl flaps were used to control the cylinder head temperatures within the required ranges. Fourteen such flaps were located about the circumference at the rear edge of each cowling. They were hydraulically operated, opening and closing to the degree required to allow the right amount of air to enter the front of the cowl, remove heat from the finned cylinders, then exit from inside the cowl into the slip-stream. The cowl flaps operated a little like the damper on a pot-bellied stove. A small hydraulic cylinder controlled one flap, with the other 13 fastened one to another by bellcrank linkages. By moving the one flap, all the others were moved a like amount. Two control levers, one for each engine, were located on the pilot's control pedestal. Using them, it was possible for the pilot to control cylinder head temperature by adjusting the cowl flap openings as required.

The carburetor airscoops were located atop the cowlings. Air coming through the scoop was directed through a large rectangular filter about 15 by 30 inches, then to the carburetor intake. The filter had to be washed in gasoline on a regular basis to keep abrasive material out of the engines—particularly when the aircraft were used off the North African desert or unimproved jungle fields in the Pacific.

Almost looking like a model builder's diorama, this J gets some field maintenance in the South Pacific. (USAF)

One of the last B-25J's to be painted olive drab. This aircraft paid a slight penalty in speed because of increased weight over succeeding polished aluminum J's. (USAF)

One last feature of the cowl was the small fairings that covered the Clayton short stub exhaust stacks on each cylinder. The short stacks were used in place of the standard exhaust collector ring and exhaust pipes because they made the exhaust less visible at night. They did have the drawback of stacks vibrating loose and, in some cases, blowing completely off in the event of a backfire. They were, therefore, a regular maintenance item. Another drawback was the much louder engine exhaust noise unmuffled by a collector ring and exhaust pipe system.

Two-speed turbo-superchargers were fitted to the engines, with a low blower ratio of 7.06:1 and a high blower ratio 10.06:1. These were used, of course, to pack increased air into the engine for far more horsepower at low altitude and to have enough air to run them efficiently at high altitude. The low-speed blowers were used at takeoff and below 11,000 feet, or even at 13,000 feet for the long-range cruising mode. High blowers were used when speed was of importance above 11,000 feet. Supercharger controls were also on the pilot's control pedestal.

The front of the engine crankcase was filled with reduction gearing which provided a propeller-shaft to crankshaft-speed ratio of .562:1. This allowed the engine to run at high revolutions per minute to develop horsepower, yet kept the propeller turning about half that speed so the blade tips would not exceed the speed of sound, a condition resulting in greatly decreased propeller efficiency. The engine was attached to a structural tubing mounting ring with Lord rubber engine mounts to absorb vibration. Without such mounts the engines would have shaken and damaged the airplane with structural fatigue in a

Close-up of Wright Cyclone R-2600-13 engine, a 14-cylinder powerplant producing 1,700 horsepower at takeoff. A Holley carburetor and air filter was fitted. The Hamilton Standard propeller hub was controllable pitch with full feathering capability. (Mendenhall)

Close-up view showing cowl flap push rod control, Clayton exhaust stacks, ADF antenna, and landing gear. Starter gear box storage is located just above main gear strut. (USAF)

relatively short time—to say nothing of the crew's discomfort. The engine mounting ring was attached to eight tubular steel legs which formed four V's from the ring attachment points to the front wing spar and nacelle firewall.

The propeller was a three-bladed Hamilton Standard controllable-pitch type with a diameter of 12 feet 7 inches. The hub was a Hydramatic full feathering unit. The 6359-18 blade type had a fairly wide chord, almost to the tip to better absorb the horsepower at relatively low air-speeds. Pitch settings ranged from fine, or low pitch of 22 degrees, to coarse, or high pitch of 90 degrees—the fully feathered position. Propeller controls were located next to the throttles on the forward pilot's pedestal. By varying the propeller pitch and throttle position, the pilot could maintain the engine rpm's required for various parts of the flight regime—takeoff, climb, cruise, etc. In case of engine failure, the propeller could be full feathered to eliminate drag induced by windmilling on the bad engine side. Ground clearance for the props on the B-25 was 11½ to 12 inches, compared to 9 inches on the B-26. This gave the B-25 the edge on unimproved runways and was one reason for its popularity in the Pacific.

A great deal of gasoline was required to operate the engines: over fifty gallons could be used just warming up, taking off and climbing to 5,000 feet with a heavily laden ship. Seven tanks, all self sealing, were built into the wing center section and upper fuselage. Among them they could hold 1,189 gallons. There were two tanks in the forward wing center section holding 184 gallons each, two tanks directly aft of them holding 151 gallons each, two 152 gallon tanks over each engine nacelle and one 215 gallon tank in the upper fuselage center section. For long ranges needed in ferrying, an additional 335-gallon drop tank could be fitted into the bomb bay along with two more 50-gallon tanks in the waist gunner's compartment. Under those conditions a potential of ten fuel tanks existed with a total of 1,624 gallons of fuel possible. Cross feeding between the tanks was accomplished by means of several auxiliary electric fuel transfer pumps controlled by switches and manual valves.

Lubrication oil required for the engines was contained in two 37½ gallon tanks, one in each nacelle. The oil systems were independent and self-sealing, and along with each tank included an engine-driven oil pump, an oil dilution system, temperature and pressure indicators, and regulators. Oil temperatures were maintained between 185°F to 205°F. Temperature was governed by two oil coolers mounted just outboard of each nacelle in the wings. An intake slot in the wing leading edges admitted cold air, the amount being regulated by pilot-controlled shutters which were downstream from the oil cooler. The shutters were located near the cooling air-stream exits on the wing upper surface near the trailing edge.

Hydraulically retractable tricycle landing gear was relatively new and in vogue at the time the B-25's lines were laid down. A look at the nose gear on most tricycle-geared aircraft reveals that the gear swings straight back about its pivot point and upward into the wheel well. Not so on the B-25; it swung back at an angle, coming to rest in a canted position in the starboard side of the fuselage nose, which left room on the port side for the bombardier's passageway to his station. This action was achieved by simply angling the gear mounting shaft. The nose wheel, capable of swiveling on the ground for directional control, had to be straight ahead for retraction; therefore, a centering device was affixed to the strut that operated automatically when the strut was fully extended with the nose wheel off the ground. In addition to the oleo shock strut, a shimmy damper was included in the shock strut unit made by Bendix. The wheel itself was made by Hayes and the tire was a 30-inch smooth contour type. There was no brake on the nose wheel.

The main gear hydraulically retracted rearward into the engine nacelles. Forty-seven-inch smooth contour tires were mounted on Goodyear wheels which were fitted with hydraulic-pneumatic brakes. The tread between the main wheels was 19 feet 4 inches. Again, Bendix manufactured the air-oil shock struts. The retraction control was located between the pilots' seats on the cockpit floor. Other related items were indicator lights showing the gear position and whether it was locked in that position, plus a warning horn that sounded when air speed got near the stall point with the gear up. The landing gear was not to be extended above 170 miles per hour.

In case of hydraulic failure, a mechanical means was available to lower the tricycle gear. This consisted of a hand crank in the radio operator's area—during a hydraulics-out landing he was a busy fellow, for it will be recalled that the emergency flap-lowering mechanism was also in that compartment. Once on the ground, braking still had to be accomplished without the hydraulic system. There was an emergency braking system fed by a compressed air bottle stowed in the upper turret compartment. An air valve, reached by either the pilot or copilot, fed compressed air through an automatic transfer valve to each wheel brake cylinder. Only one landing could be made before the air bottle had to be recharged. Such landings were precarious, as the pilot could not apply braking to each wheel selectively; therefore, engine power had to be used to straighten the ship out. That was okay, but adding power when wanting to stop was somewhat at cross purposes.

The bombardier had the aircraft nose section to himself. It was reached by entering the front hatch and crawling under the pilots' floor into the greenhouse. Once in, it was reasonably roomy. Just behind the flat-glass bomb

This close-up front view of a B-25J shows the "double" package guns being wrapped in cellulose material prior to flight to a combat zone. Also of interest are the two .50-caliber fixed guns on the bombardier's compartment right side to complement his flexible .50. (USAF)

aimer's panel was a universal mount to take Norden, Estoppey or British Mark IX sights. The left wall, just behind the transparent nose, held a fairly complex instrument panel containing, among other things, an outside air temperature gauge, altimeter, air speed gauge, bomb drop interval control, a bomb release handle, bomb-bay door controls, salvo switch, bomb rack selector switch, bomb release switch and a selector train switch. In addition, there were a bombsight defroster tube, various light switches, pilots' call switch, and heater controls for a heat register located just aft of the control panel. This constituted most of the bomb dropping equipment.

On the right side of the compartment, along with oxygen regulators and phone jack, were two .50-caliber Brownings, one fixed and one hand-held. In some cases, both were fixed and were fired by the pilot through a system of bungee cords linking the charging and firing mechanism with the cockpit. Each gun carried three hundred rounds of ammunition. On some of the later Model B-25J's, in addition to two fixed guns, there was a third hand-held .50 which had an additional two hundred rounds. At that point, the bombardier's nose compartment was getting to be more like an artillery battery! In addition to the guns and ammunition boxes, there were canvas containers for ejected shells. Against the right side (at the compartment rear) was a riding seat for the bombardier. The seat and back were armor-plated on early B-25's; the whole rear wall and floor was armor-plated on the J model. A bicycle-type seat, mounted on a pedestal, was located further forward in the middle of the compartment floor for use while aiming during actual bombing runs.

The bomb bay, controlled from the bombardier's position, was nearly ten feet long and full-fuselage width. It extended almost to the top of the fuselage—interrupted only by the narrow passageway from the top turret compartment to the waist gunner's position. The bay, as mentioned, could carry a droppable 335-gallon auxiliary fuel tank, or, if required, target towing equipment. But its real use was bomb storage. Maximum bomb load of the B-25J was 4,000 pounds; the maximum size bomb it could carry was 2,000 pounds. However, only one 2,000 pounder could be carried due to the sheer size of the bomb. Combinations of other bombs were: two 1,600-pound armor-piercing; three 1,000-pound general purpose; eight 500-pound semi-armor-piercing; eight 250-pound semi-armor-piercing; or six 500-pound general purpose bombs. The difference between armor-piercing, semi-armor-piercing and general purpose bombs was the ratio of explosive-charge to weight for the bombs. The ratio ran something like 51 percent explosive for general-purpose bombs, down to only 14 percent for armor-piercing, with about 31 percent for semi-armor-piercing.

While bomb-bay door opening and bomb dropping was normally controlled from the bombardier's compartment, the pilot could, in an emergency, salvo the bombs (all bombs *out at once* regardless of target requirements). Also, if the hydraulically opened doors were inoperable they could be cranked open from the navigator's compartment and the bombs or auxiliary tank dropped by means of a switch within the bomb bay.

The flight deck, or pilots' compartment, contained the pilot and copilot seats, both covered with 145 pounds of armor plate and mounted side by side. The cockpit floor under the seats was also covered with armor plate. The instrument panel with its 30-plus dials and switches had three major classifications of instruments—flight attitude, navigation, and engine monitoring—and then there were the special instruments because the aircraft was a bomber. There was not a duplication of instrumentation on the B-25 for pilot and copilot; instead, the flight attitude and navigation instruments tended to be located on the pilot's side while the powerplant and system related gauges were on the copilot's side.

Since the width of the entire panel was only about four feet, it was possible for either pilot to view the entire instrument array and fly the aircraft safely from either seat.

PILOT

Airspeed indicator
Directional gyro
Gyro horizon indicator
Pilot direction indicator
Clock
Remote reading compass
Turn and bank indicator
Altimeter
Turn indicator
Rate of climb indicator
Bomb release signal light
Bomb doors open signal
Radio compass indicator
Compass card holder
Glide path indicator
Gunsight rheostat
Static pressure selector
Suction gauge

COPILOT

Manifold pressure indicator
Oil pressure indicator
Tachometer
Fuel pressure indicator
Auxiliary tank fuel level
Landing gear signal light
Nose wheel turn indicator lights
Hydraulic pressure gauge
Cylinder temperature indicator
Oil temperature indicator
Free air temperature indicator
Carburetor air temperature
Main fuel tank level indicator
Landing gear & flap position
Brake pressure gauge
Rear wing tank fuel level
Front wing tank fuel level

Below the instrument panel was an electrical switch panel for formation lights, identification lights, interior fluorescent lights, battery disconnects, etc. On the copilot's side, the small sub-panel contained a pull-out extension maplight and engine compartment fire extinguisher selector. In the center of the panel and extending outward into the cockpit was the pilot's control pedestal containing engine and propeller controls. On the floor between the seats was the emergency hydraulic system hand pump and emergency compressed air control for the brakes. The cockpit sides were given to oxygen regulators, head-set plug-ins, heater controls, jack boxes, side window latches and a few cockpit radio controls. The normal dual wheel control columns and rudder pedals with toe brakes completed the cockpit.

Behind the cockpit, in what had previously been the navigator's compartment (in the B-25A through G), there was now the top turret. The Bendix Model R turret sat on a tubular pedestal that mounted to the bomber's floor. It contained two Browning M-2 .50-caliber machine guns located side by side with the gunsight between them. The gunner sat on a bicycle seat that jutted out from the turret, his feet resting on short bars that projected from the pedestal on either side. These foot rests were adjustable in height. The turret had an armor-plate apron that extended from the plexiglass dome to about the height of the seat and for about 180 degrees, covering the gunner's body from a head-on attack. The guns were hydraulically charged and fired with electric trigger solenoids and each had four hundred rounds of ammunition.

View of the pilot's side of the control panel with mainly flight situation instruments and operational switches for his attention. (USAF)

One-half-horsepower electric motors drove the turret, rotating it at 33 degrees per second horizontally through 360 degrees and through elevation from horizontal to -92 degrees. In the event of electrical failure, the guns could not be fired manually, nor could they be fired in any position in which they were trained on parts of the airframe or propellers. To fire the guns, a charger switch was held down until the gun bolts were moved to the rear position. The master control handle was then gripped and rotated to the neutral position and the main power switch at the back of the handle depressed. Moving the handle up and down, the guns followed up and down, and by twisting the handle the turret rotated. After lining up the target in the gunsight's cross hairs, the trigger switch was depressed on the control handle. In addition to the guns, an oxygen outlet and a radio jack box for communications were provided in the turret for the gunner.

The copilot's panel mainly consisted of instruments for the engines. Panel was narrow enough that either could view the other's instrumentation. (USAF)

When facing forward, the upper turret could bring to play, in conjunction with the two fixed nose guns plus two forward firing package guns on either side of the fuselage, eight guns—all .50 caliber—on a strafing target. The package guns were mounted in "blisters" on the lower fuselage just back of the pilots' canopy. Each of the four had four hundred rounds of ammunition. A gunsight mounted over the instrument panel on the pilot's side enabled him to aim and fire the nose guns and package guns at the same time with a button on the control wheel.

The turret area also housed the radio equipment, for which the turret gunner usually served as radio operator. Equipment included a command set, a liaison set, an emergency transmitter, a radio compass, a marker beacon, a contactor unit and interphones. Other sundries included a transmitter key, small table, thermos bottle, oxygen regulator, ashtray and relief tube.

At ceiling height was an opening in the turret compartment rear wall that was surrounded with ¼-inch armor plate. It was designed to crawl through and along the top of the bomb bay to the waist gunner's position, a compartment that contained .50-caliber flexible guns mounted to the rear of each side bay window. Each waist gun was provided with 250 rounds. Below the guns was a case and link ejection bag to catch the spent ammunition. This was necessary for two reasons. The first was to keep the spent cartridges off the floor—standing on them would be like standing on ball bearings, hardly conducive to gunners keeping their footing in a wildly lurching aircraft in full combat. The second reason had to do with the alternative of ejecting the shells overboard, which raised the good possibility of dropping them on aircraft fly-

The center control console contained controls easily reached by either pilot—flaps, cowl flaps, throttles, propellers, landing gear and assorted master switches. (USAF)

ing in formation below and resultant broken windshields or much, much worse. Also in this compartment, on the front wall was a fire extinguisher and on the rear wall a combustion-type heater along with the magnetic compass head.

Moving rearward through the fuselage from the waist gunner's position there was the photographer's station with suitable ports and mountings for a Fairchild aerial camera. It included a vacuum valve and junction box for powering the camera and an intervalometer for controlling it. A chemical toilet was located on the floor between the waist guns and photographer's area. Also in this compartment was an alarm bell (pilot's warning for a bailout) and an interphone jack box. Armor plate was on either side of the doorway leading to the waist gunner's position.

Ammunition boxes, holding six hundred .50-caliber rounds each, were located on either wall of the photo compartment with tracks extending from them to the Bell M-7 tail turret with its two .50-caliber Brownings. The ammunition belts ran along a track on the fuselage side walls from the ammo boxes to the gun breeches seven or eight feet away.

The tail gunner sat upright on a bicycle-type seat, his head in a canopy where he had almost unrestricted 360-degree vision. From his shoulders down, armor plate protected his body against any attack from the six o'clock area. A type M-8A optical gunsight was mounted on a support above the guns at eye level. The guns themselves were trained remotely by an electro-hydraulic system controlled by two handles, one for azimuth and one for elevation. A large crossbar was also attached to the guns so that they could be maneuvered manually by the gunner in the event of hydraulic failure.

Also interesting were the systems required to combat cold and icing as were often encountered in winter operations in Europe and the Aleutian Islands off Alaska. Ice, long the nemesis of piston engine aircraft, tends to build up on the wings and control surfaces under certain conditions, somewhat like those encountered on the ground with freezing rain. As a result, two things happen simultaneously. The wings' airfoil sections become distorted due to the frozen buildup, thus creating less lift at a given airspeed, and the total weight of the

The tail gunner's position in a B-25J wasn't all bad. He could sit upright, he was protected by a lot of armor plate and the turret had two .50-caliber power operated guns with 600 rounds per gun. (USAF)

While not a "J" model, this photo shows one other capability of most B-25's near the war's end: Four Jato bottles propel this aircraft into the air in less than half the normal takeoff run. (USAF)

aircraft also increases drastically due to the entire airframe getting covered. In the extreme, the excessive weight and lack of lift result in a stall condition and a lead sled ride to the ground.

To prevent this, deicer boots were provided on the wing and the tail. Manufactured by B. F. Goodrich, they consisted of heavy rubber sheets wrapped around the leading edges of the flying surfaces. Beneath the sheets were collapsed rubber tubes, as many as six on some sections of the wing and one on each of the tail surfaces. By a system of valves air pressure was automatically directed to the tubes, enlarging them and then contracting them as they bled down. That caused the rubber sheet surface to undulate, breaking away the ice in chunks and allowing the high-speed slipstream to carry them away behind the wing and control surfaces. It generally worked very well. The airfoil section was maintained and, of course, so was maximum lift. In a very few cases, though, an insidious type of rime ice was sometimes encountered that was more or less slushy. In this event, the pulsating of the deicer boot would not crack the ice but merely move it out of the way of the deicer where it would continue to build up. When that happened, only a fast descent to a lower altitude and hopefully warmer temperatures could save the aircraft. Climbing up to warmer temperatures was also a possibility but that solution was usually unrealistic because of the lack of airfoil lift.

Another portion of the aircraft that had a real problem with ice was the propellers. The blades were airfoil shaped in cross section for maximum efficiency in delivering forward thrust. With ice buildup on them, they became nothing more than whirling clubs flailing away in the air. That added to the problem of maintaining airspeed along with the previously mentioned weight gain of the aircraft and deterioration of the wing airfoil. No thrust, overweight, and no lift—the three witches of MacBeth in spades!

The answer to the propeller ice problem was coped with in a different way: deicing liquid rather than flight surface undulations. A 4½-gallon tank filled with a solution of 85 percent denatured alcohol and 15 percent glycerine did the trick. On cockpit demand, the ice-melting liquid would trickle out of the deicer slinger tubes on the propeller hub, run out the blades due to centrifugal force, and melt the ice loose from the blades. This it did, but sometimes the ice came off in chunks hurled by the velocity of the propeller and some of the chunks would fly against the fuselage sides with great force—inside it sounded as if the ship had received a direct hit from a German 88. Also, some of the larger chunks left significant dents in the fuselage sides. Other lesser combatants of Mother Nature's icebox were the hot-air windscreen defroster and the gasoline-fired space heater, its heat directed to various parts of the aircraft—just like in a house. One last shot at the cold was a draw curtain at the rear of the cockpit which prevented cold drafts from coursing up and down the length of the ship.

The B-25 was a sophisticated product of the late 1930's aero engineering and manufacturing technology. It represented an era when aircraft designing was at last based on resultants of mathematical formulas and tested scientific principles rather than cut, try and hope. Aircraft such as the B-25 formed a significant part of the solid scientific base from which the computer-fired high-technology jet age developed.

After the war, the "J" soldiered on for many years. Its turrets and armaments were removed—the new jets were the combat machines of that time. However, the 25J, defanged, was still a great airplane for multi-engine, fire control and navigational training as these four views of a typical example show. (USAF)

4

MISSION FROM SHANGRI-LA—AND OTHERS

IT WAS NOT WHAT YOU would call a great day for flying. Standing on the 809-foot flight deck of the recently commissioned *USS Hornet* was physically uncomfortable at best. The leaden skies spat rain squalls from time to time and quick cold gusts made sure some of the dampness penetrated the outer clothing to the skin. The twenty-thousand-ton ship rolled easily in the swell of gray-green ocean expanse, but on occasion a wave did break over the 83-foot-wide flight deck as the ship's 120,000-horsepower turbines pushed her along. The spume from breaking wave tops shattered into white lace against the dark sea, then disappeared as bubbles burst and once more became only sea water. Patches of low-flying scud clouds and fog completed the bleak seascape.

During the very early morning of April 22, 1942, the *Hornet* was 824 miles from Japan—its goal, a position 650 miles off the coast of the island aggressor. At 3:05 A.M. the situation changed rapidly. Radar had picked up a hidden stranger to the west, a patrolling Japanese picket boat. Fifty of these small vessels had been staked out by the Japanese in a 650-mile perimeter around their home islands, just in case the Americans had any ideas of attacking them. The little boats had even picked up bits of radio communication between the *Hornet* and Admiral Halsey's eight-ship task force, including the carrier *Enterprise,* which was to accompany the *Hornet* to the Japanese doorstep. The picket sent a warning radio message to the homeland and the cruiser *Nashville* of the U.S. task force set out at once to quickly sink the little vessel. Lieutenant Colonel James H. Doolittle was ordered to get his planes in the air at once. There was little time to spare as the now-warned enemy would quickly be sending bombing squadrons to sink the U.S. task force to the bottom of the gray Pacific.

Almost five months before, the mission had begun on January 4, 1942. Only a few weeks earlier, the United States had felt the full brunt of Pearl Harbor and the ensuing onslaught on the Philippines. Seemingly nothing could stop the Japanese tidal wave as it began to fan out from the home islands, its goal to make the Pacific a Japanese lake. The American public was rapidly becoming disheartened, never having dreamed a little nation like Japan could suddenly kick this country in the behind and make it stick—and, as they broadly advanced, keep on making it stick.

Through a round of inventive thinking at the request of Admiral Ernest J. King, a subordinate, Captain Frances L. Low, suggested that if an Army bomber with its superior bomb load could take off a carrier, then perhaps a carrier attack force could approach the Japanese mainland and give them a good dose of their own medicine. The top brass liked the idea. If they could run those bombers in close enough to Japan on a carrier, they could take off, bomb their targets and continue on to the safety of the Chinese mainland. Admiral King sought a recommendation for what type bomber to use. The B-25 was thought to be a good possibility.

With windswept seas in the background, the starter (flag in hand on right) gives the readiness signal to a Tokyo-bound B-25 on the USS Hornet's flight deck. (USAF)

Deck crew ducks as one of Doolittle's B-25's starts to roll on Tokyo bombing raid. (USAF)

The USS Hornet, decks laden with Tokyo-bound B-25's, plows through heavy seas as one of the first of the mission's 25's takes off and climbs out. (USAF)

This information was passed on to USAAF Chief of Staff Henry Harley "Hap" Arnold who immediately conferred with Jimmy Doolittle on the project. After starting off in the Air Corps in the 1920's, Doolittle had left the service and carved out a swath of heroics through the Golden Age of the 1930's, primarily as a fantastic air race and stunt pilot, but also as a first-line executive with the Shell Oil Company. With the advent of U.S. involvement in war he had returned to the service as a lieutenant colonel. He studied the bomber problem and gave the following critique to Arnold, who had asked, "What airplanes have we got that will take off in 500 feet, with 2,000 pounds of bombs and fly 2,000 miles?" Doolittle's response:

1. The Douglas B-18 did not have the required range.
2. The Douglas B-23 was too big in span. It would never clear a carrier's superstructure.
3. The B-26 was OK on everything except take-off distance. It needed over 1,000 feet to get off and airborne—longer than a carrier deck by a good margin.
4. The North American B-25 could do it.

The B-25 was the aircraft chosen for the mission. Calculations showed the new carrier *Hornet* could hold sixteen of these bombers and carry them within 450 miles of their target.

The B-25 had become operational by this time with the 34th, 37th and 95th Squadrons of the 17th Bomb Group of Pendleton, Oregon. The group had previously been stationed at McChord Field and had operated the bulky Douglas B-18's. The 89th Recon Squadron was also attached to them. The group had already had a slight bit of activity against the enemy by flying anti-sub patrol off the northern Pacific coast. They were given credit for having deep-sixed an enemy submarine during this period.

Now they were scheduled to move to Columbia, South Carolina, in February 1942. Doolittle got to them. He contacted all B-25 squadron leaders with a call for volunteers for a very hazardous mission. With patriotic fervor soaring high right after the Japanese attack, the volunteer response was almost one hundred percent from the Air Corps personnel approached. The volunteers quickly reported to Eglin Field in Florida to train for their super secret mission. Meanwhile, their B-25B's were being modified to help them carry out the mission. The main thrust of the rework would be more tanks for longer range and removal of armament for lighter weight.

The crews were also in for a rework. In their case it was learning to take off in a heavy condition from the short length of a carrier's deck. A typical "deck" was marked in outline on the runway and, with a carrier-experienced Navy instructor, the crews learned to open up the Cyclones, with brakes locked, until their Mitchell seemed ready to literally tear itself apart. At brake release the B-25 was hauled off at the earliest possible moment to stagger into the air, then gain speed for a normal climb out. (This became known by all as a "Hornet takeoff" and was performed from time to time, just for the hell of it, by most B-25 pilots during the remainder of the war.) Finally they were ready, right down to the black-painted broomsticks mounted in the tail observers' bubbles to simulate machine guns. That was Doolittle's suggestion. The pilots and crews were now razor sharp in short-field (carrier deck) takeoff techniques.

With all of this preparation going on, someone asked a very obvious question: In the real world, could a heavily laden B-25B actually make it off a carrier deck and not drop into the sea? To find out, two B-25's were hoisted aboard the new *Hornet* at Norfolk for a short journey out of sight of the prying eyes of land. Though the *Hornet*'s deck was long, there were all sorts of cranes, aircraft and other objects using up its rear portion. Only about 450 feet were left for the actual takeoff of the B-25. One good thing: headed into the wind the airspeed indicators on the Mitchells registered 45 miles per hour while standing still on the carrier deck when the ship speeded up into the wind. The number one B-25, with the built-in 45-mph headwind, locked brakes, went full throttle and let go. Its pilot, Lieutenant John C. Fitzgerald, with over 400 hours in the B-25 type, was amazed at the rapid aircraft jump-off from the flight deck. He quickly gained altitude and watched his partner in the test, Lieutenant James F. McCarthy, follow suit. They happily joined up and flew back to Norfolk. B-25's *could* be flown off carrier decks.

Since the proposed mission—bombing Tokyo and other cities—was to be low level, the Norden bombsight was discarded and a crude Kentucky windage sight substituted. It cost about twenty cents to make and was called a "Mark Twain" sight. Tokyo targets would include an army arsenal and an armory, plus some steel, chemical and gas plants along with docks and refineries.

April 1 found the *Hornet*, after a hasty trip through the Panama Canal, sailing westward through San Francisco's Golden Gate. Sixteen B-25B's were moored to the decks and their crews plus aircraft spares were put up below deck. The task force was under the command of Captain Marc A. Mitscher. The westward trek was to take two weeks and the flight crews, now fully aware of their destination, grew more uptight with each passing day.

Now suddenly the picket had discovered them and there was no longer any time to lose. With advance information the enemy would make its welcoming party an utter disaster for the U.S. bombers. They had to go at once. By 8:20 A.M. Doolittle, in the lead ship, was off. He circled the *Hornet* once, then headed westward toward Japan. Within an hour all sixteen aircraft were off the deck and likewise heading for their targets. With their charges gone, the carrier force reversed its course and got out of range of the Japanese Air Force.

The Doolittle Raiders now hurried on along their courses to targets in Japan, dodging picket ships or Japanese patrol bombers here and there. The cloudy, dark weather improved as they sped toward their targets, and just before the Japanese islands they burst into the yellow sunshine of an April day—their goal dead ahead. Doolittle and company, though their bomb loads were relatively light, did damage enough to make the enemy sit up and take notice. The Japanese were forced to believe that they could not divert all their fighter aircraft to far-flung Pacific combat. Some would have to remain at home and watch the islands, and that hurt. They were already spread so thin with their combat excursions, the effect of retaining planes at home would be severely felt on some battle fronts.

Their targets hit, the bombers headed for the Chinese mainland, except for one. Low on fuel, it headed for Vladivostok, Russia. Landing there, the crew was interned, for Russia was not at war with Japan. The other fifteen B-25's continued heading for China and soon ran into foul weather again—and then, darkness. All crash landed on or near the China coast. Thirteen crews made it to the friendly Chinese and eventual safety. Two crews, however, fell victim to Japanese patrols and later torture—in two cases, death.

Doolittle was crushed. Eighteen planes were expended along with the deaths and captures amongst the men he had led. America felt otherwise. Newspapers proclaimed "Tokyo bombed!" "Doolittle Dood It!" the headlines blared. It was front page stuff for a United States that had read of a daily drubbing at the hands of the yellow tide sweeping the Southwest Pacific. Roosevelt said the bombers had come from Shangri-La, a mythical Tibetan kingdom of the then-popular book and movie *Lost Horizons.* He did this to keep the location of the Navy's carriers secret from the Japanese. As General Doolittle put it during an interview in *Shell News* in recent years, "It caused the Japanese people to question their warlords, who had told them Japan would never be bombed; it gave our people at home the first good news since the war started—and it caused the Japanese to retain for the defense of the home islands aircraft that would have been much more useful down in the South Pacific, where the battle was going on."

For his great efforts Doolittle was presented the Congressional Medal of Honor by President Roosevelt. He quietly accepted it, with his wife, General Arnold and General Marshall standing by. He went on to command the

Twelfth Air Force in North Africa and then the mighty Eighth Air Force in Europe. By war's end he had been promoted to lieutenant general—a fantastically high rank for a reserve officer.

While the Tokyo raid was, without doubt, the most famous exploit of the B-25, it was but one of many contributions to the winning of World War II. As mentioned earlier, the 17th Bomb Group had traveled to Columbia, South Carolina, from Oregon, for the German U-boats were becoming a real menace along the Atlantic Coast. At times the submarine wolf packs were sinking ships within sight of the shoreline. Coastal patrol bombers were an answer. Two other B-25 Bomb Groups, the 13th and 21st, joined the 17th in this work.

Before the end of 1942 the sub problem had declined and the 17th switched over to the B-26. The B-26 was now becoming available in quantity and the seasoned 17th crews transitioned quickly and were off to North Africa. The 13th Bomb Group was inactivated and by the end of 1943 the 21st was also inactivated. Two other B-25 Bomb Groups were formed for training purposes, the 309th and 334th, and spent the war in the U.S.

With a red and black "Thunderbird" insignia on a white background emblazoned on the fuselage and blue cowl nose rings, this B-25A belongs to the 17th Bomb Group, the first Army Air Corps operational unit outfitted with that aircraft. Note the prone tail gunner's position. (USAF)

A B-25C climbs out on a training flight. Note lack of guns in either dorsal turret or nose. (USAF)

"Lady Jane" in anti-submarine patrol paint job was a B-25C. Forward and rear crew entry hatches show up well in this photo. (USAF)

A Netherlands/East Indies flight of B-25C-5-NA's on patrol. Ships were formerly RAAF. (USAF)

This unarmed B-25 sits out a sunny day that hopefully will find it airborne before evening comes. (USAF)

This photo shows the tail observation bubble in detail as used on B-25C's and D's. The Douglas A-20 seems a stranger in the line-up. (USAF)

The first actual B-25 raid on the Japanese had preceeded Doolittle's mission by a week. It was carried out by the 3rd Bomb Group of the 5th Air Force, on the long-range mission from Australia led by Brigadier General Ralph Royce and Colonel Paul "Pappy" Gunn. All through 1942 the 13th and 90th squadrons of the 3rd Bomb Group interdicted in New Guinea, with the noses of their B-25's carrying ever-heavier fire power under the impetus of Colonel Gunn's urging. In August the 38th Bomb Group—also B-25's—joined them. The strafing and low-altitude fragmentation bombs worked so well that 175 of the B-25C's and B-25D's were converted for this mission at the Townsville, Australia, maintenance depot.

The Battle of the Bismark Sea, New Britain, Leyte, Papua and the enemy airstrips at Rabaul gathered all kinds of attention from the rampaging B-25's during 1943. The 5th Air Force had been further strengthened with the 345th Bomb Group, the colorful "Air Apaches" in June. The conversion of the 22nd to Mitchells, after their B-26's were pulled out of the Pacific for duties more suitable for them in the European theater of operations (ETO), added to the low-level attack strength.

The addition of B-25's to the 13th Air Force in the form of the 42nd Bomb Group in June 1943 and the 41st Bomb Group in December 1943 rounded out the Pacific battle contingent. By the end of 1943 the island campaign was definitely going on the side of the U.S. and, before too many more months passed, B-29's would start to upend the landscape of the Japanese mainland cities.

Of course the war was global and the B-25 saw service in other areas as well. The 10th Air Force of the China-Burma-India (CBI) theater had two squadrons of B-25's in its 7th Bomb Group, the 7th and 11th. The 7th went to China in June 1942 and the 22nd was stationed in India, eventually forming a nucleus

This B-25C is fitted with deicer boots on wings and tail. Photo also shows dorsal turret guns raised to maximum vertical position. (USAF)

These Mitchells lay it on an oil storage dump at Laichokok, on the Chinese mainland opposite Hong Kong. (USAF)

Puffy cumulus, and between the cloud breaks a lush green countryside below—what a day for flying, particularly a docile B-25 on a training flight. (USAF)

An F-10 (later RB-25) tries out its tri-metragon camera high above the Colorado Rockies. (USAF)

Mitchells from the 38th Bomb Group careen low over a Japanese ship in the Sorong area, Dutch New Guinea on June 17, 1944. (USAF)

A 9,000-foot-long bridge over the Yellow River should make a large target; however, this 14th Air Force B-25J doesn't seem to be doing too well. The flak tower in the foreground doesn't help much. (USAF)

This B-25D patrols low over a U.S. destroyer near the Alaskan coastline. (USAF)

This B-25C "El Diablo IV" is in the process of softening up New Britian Island prior to coordinated land, sea and air attacks on Cape Gloucester. (USAF)

With the landing forces for Cape Gloucester below, these B-25C's cover the area. Note the painted-over bombardier's compartment glass and the multi .50-caliber machine guns in nose—probably installed under Pappy Gunn's direction in Australia. (USAF)

The barren North African desert is the background for these Mitchells (B-25C's) heading for their targets. (USAF)

for a new 341st Bomb Group in September. The 11th Squadron concentrated on raiding Formosa; it occupied mainland China along with the Chinese-American Composite Wing's 1st Bomb Group—also B-25's—which was formed in October 1943. The 341st Group became the famous "Burma Bridge Busters" after learning the trick of low-level placement of their bombs into those sturdy structures.

Camouflage for the aircraft in the Far East was generally olive drab on the upper surfaces with pale gray undersides. In general, unit markings were fairly subdued, with perhaps the exception of the 5th Air Force "Air Apache" squadrons with the exotic bats and other faces painted on the noses of their strafers along with colorful Indian heads in full headdress emblazoned on the vertical tail surfaces. In the CBI, some squadrons took to painting white barber pole stripes around their fuselages along with the usual assortment of nude and semi-nude pinup girls on the fuselage sides under the cockpit windows. Of course, as the war progressed and the enemy airforce attacks became more scarce, the o.d. paint was no longer provided and a polished aluminum natural finish became standard.

On March 1, 1943, the U.S. Marines at the Cherry Point NAS in North Carolina formed a combat squadron of B-25's (or PBJ-1's, the Navy designation). It was called VMB-413, with -423, -433, -443, -611, -612 and -613 squadrons being formed rapidly thereafter. They were soon in combat, a total of 105 aircraft (of which 45 were lost before the war was over). The B-25H cannon carrying models were used and the J's were equipped in many cases with radar pods, either under the nose or on the wingtips. The radar planes were to lead night attacks on the enemy, an excellent case in point being the night rocket attacks off Saipan in 1945. The aircraft were painted in three shades of blue: dark on top, lighter on the sides and pale-to-white on the undersides.

The 7th Air Force 41st Bomb Group (B-25's) had run the gamut of Hawaii, Tarawa, Eniwetok, Makin and back to Hawaii. There they picked up the new rocket-carrying B-25's which had supplanted the cannon models. They soon re-

This base in North Africa had belonged to the Germans. Now B-25's of the 9th Air Force drift in for a landing there. (USAF)

"Tokyo Sleeper" appears in great shape as it awaits a mission with the 38th Bomb Group at Port Moresby, Papua, New Guinea. (USAF)

turned to combat flying out of Okinawa, harassing Japanese shipping and tactical targets in and around the Marshall and Carolina island chains.

While still on that side of the North American continent, it should be mentioned that the sturdy B-25 also saw action in the Alaskan 11th Air Force, which had obtained its B-25's in December 1942. During the Aleutian chain campaign the island of Attu was recaptured from the Japanese and soon acted as a springboard for the second visit of B-25's to the Land of the Rising Sun. This time it was a raid on Paramushiro on July 10, 1943. They further harassed the Kuriles between April 1944 and war's end in August 1945.

Moving the calendar back a bit to early 1942 in the North African and Mediterranean campaigns, it becomes apparent the Mitchells also contributed there. In July 1942, under the command of the 9th Air Force, the 12th Bombardment Group landed in Egypt. They flew there the hard way. Starting in Florida, the route took them on navigational legs to the Antilles, Brazil, Ascension Island and Central Africa, then on to Egypt—with no loss of aircraft. They soon learned about aerial combat by joining in with RAF formations against Axis airstrips and port installations that served Rommel's war machine. They flew ninety-one missions and dropped a hefty 1,536,000 pounds of bombs. Forty-six B-25's went through the battle of El Alamein. Ten were lost—four on one night raid when the long plumes of exhaust flames from their Cyclones made them easy targets for the enemy. In August 1943 the group was transferred to the 12th Air Force where they remained until January 1944 when they went to the CBI theater. During the desert operation many of the B-25's were painted a pinkish sand color (sometimes mottled with olive drab) with sky-blue undersurfaces.

With the success of driving the Germans out of North Africa, three 12th Air Force Bomb Groups, the 310th, 321st and 340th, continued the pressure on the Germans with attacks on Sicily and the Italian boot itself. Joseph Heller's best selling novel *Catch 22* joins them during that campaign. They were successful—blowing bridges, sinking warships and generally making a real pain out of themselves as far as the Germans and Italians were concerned.

With all the B-25's success it would seem that the 8th Air Force in the ETO would have wanted a few squadrons of them—particularly with Doolittle heading it up. As it was, only the 25th Bomb Group, a recon outfit, ever flew operationally out of Great Britain under the stars and bars of the USAAF. Even these were intermixed with other types of observer aircraft as they flew

Brass monkey weather at Ladd Field, Alaska. It's minus 35° F and the line of B-25's don't seem eager to start. (USAF)

Oil splatters galore on this B-25C's nacelles as this full top view shows. Note deicer boots on wing and stabilizer leading edge. (USAF)

This B-25J strikes at a bridge north of Nice, France. Note that all gun positions are manned in this photo. (USAF)

ZZ
130181

the British coasts watching for enemy activity as well as spotting downed flyers coming back from raids on the Continent.

Other USAAF groups, and this does not include all that used the B-25, were the 1st, 2nd, 5th, 11th, 26th, 65th, 66th, 69th, 70th, 74th, 75th, 76th and 77th training groups (recon).

The British, during the war, had received over eight hundred Mitchells on lend-lease, encompassing three models: the Mitchell I (B-25B), Mitchell II (B-25C and D) and Mitchell III (B-25J). In general they were painted olive drab with gray undersides, sported the RAF night roundels and carried large red 'codes' on their sides. The Mitchell I's never saw combat with the RAF but were put to good use training aircrews for the type. The Mitchell II's went to the No. 180 and No. 98 Squadrons of Number 2 Group where they made their first raid on January 22, 1943, against oil refineries in Belgium. As more Mitchell II's arrived, a Dutch squadron (320th) was formed, and two more RAF units as well (No. 342 and No. 305).

The Mitchell III's were used as replacements for the II's as the latter were damaged or worn out. D-day saw black and white invasion-striped British Mitchells performing their tactical roles. They served the RAF well to the end of the war as close support for Allied armies advancing through France and Holland. In September 1944 they hit Germany for the first time in a raid on Cleve. After the war the Mosquito XVI took the B-25's place in the RAF.

Other countries used the Mitchell in World War II. Russia, with over 800 delivered by lend-lease, flew them like fighters in a tactical "free for all" role. Brazil used 29 of them for anti-sub patrol up and down her long coast line. The Netherlands obtained 249 B-25's and generally flew with USAAF units in the Pacific theater. In China, 131 B-25's had the Chinese insignia on them as they fought the Japanese and later the Communist Chinese. Many were captured by Mao's masses and were used—red stars on their sides—in the Communist cause.

Robert A. Lovett, Assistant Secretary of War for Air, summed the B-25 up well in a contemporary issue of the *Army and Navy Journal.* He said: "There are few more dramatic examples of the advantages of constant improvement of current models than the modifications and design changes in the B-25 medium bomber. It is serving in every theater of war completely around the world. A superb medium bomber, the Fifth Air Force in the Southwest Pacific, working with North American, Inc., has converted it into a devastating attack-bomber with tremendous firepower forward and a specialized technique in the bombing of shipping."

Three of the ten trimetragon F-10's built practice a recon flight over a Caribbean island. "ZZ" is named "Pistol Packin Momma."

Neither rain, nor sleet, nor erupting volcanoes could keep the 12th Air Force from attacking the Germans in the Cassino, Italy, area. Mt. Vesuvius spouts ash and smoke thousands of feet in the air. B-25's at the field near the mountain's base had their fabric-covered control surfaces burnt off by the falling hot ash as well as their runways made unusable by the six-inch ash fall. Air around the volcano was very turbulent and made for rough flying conditions. (USAF)

This B-25D has been stripped of all armament for use as a weather plane. Photo taken June 22, 1944. (USAF)

5

ANATOMY OF A MARAUDER

TALK ABOUT CLASS! The B-26 prototype fresh from the factory: a shiny, polished, streamlined aluminum projectile on wings—a love affair at first sight for all who saw it. The sleek metallic sculpture was dubbed "The Flying Torpedo" by the press and that it was. It was one of those "doing a hundred miles an hour standing on the ground" designs. If someone had asked what the ultimate twin-engined air racer would look like, the B-26 had to be it—and if you didn't know better you might even think one of the old Gee Bee designers of the early thirties had a hand in it!

Compared to the North American B-25 Mitchell, the Marauder had many features that pushed the aerodynamic state of the art out to its very limits. Since almost all of the innovations were in by the time of the B-26C, that model will be used as the basis for this chapter, an in-depth look at the aircraft's construction, dimensions, systems and performance.

In total 5,157 copies of the B-26 were produced. While it was hot (it usually landed at 130 miles an hour), it served the medium bomber role well, and when the final bean counting was done at the end of World War II, it was found to be one of the safest places to be in that conflict—if you were in a bomber crew. It had the lowest loss rate on operational missions of any combat aircraft in the ETO—less than half of one percent went belly-up. However, the pilot (or autopilot) had to fly it all the time. On autopilot it would not groove like the B-25 but was constantly climbing, descending, and yawing, always hunting for an elusive stability that was not there. Hand flown it was much the same and most pilots, if they had any flight time ahead of them to speak of, would have been delighted to transfer to the more docile B-25. Though restricted from fighter maneuvers such as loops and rolls, pilots felt it handled more like a giant fighter than a lumbering bomber.

It was designed for a big crew of seven men: bombardier, pilot, copilot, navigator, radio operator/engineer, camera operator and tail gunner. On long missions no one was overworked; in fact, there were three extra seats in the aircraft dubbed fatigue positions. For combat the radio operator took up the dorsal turret gunner's position. The aircraft's mission was to carry a sizeable bomb load over a medium range and take out a bridge, blow up an ammo dump or a rail yard in support of advancing ground forces. This was generally to be accomplished from medium to low altitudes.

The fuselage of the B-26 was absolutely circular in cross section except for two small flat areas behind the cockpit side windows that faired back to the wing's leading edge. Its largest diameter measured 92 inches and the overall length was 58 feet 2 inches. In general, the fuselage side and top views followed the proportions of the wings' symmetrical airfoil. The cockpit windshield and the round dorsal gun turret were perfectly streamlined protuberances into the airstream.

One of the 201 Martin B-26's (40-1498) with a stubby racer-like span of 67 feet—2½ feet less than the smaller, lighter B-25. (USAF)

Construction of the fuselage was somewhat different than conventional all-metal methods such as that used on the B-25. During an interview one of the prime B-26 designers, Peyton Magruder, was quoted as saying the fuselage was made up of "four stout longerons that were wrapped in heavy aluminum sheeting" and to this was added a heavy keel similar to a flying boat. That brief but basic explanation of the fuselage structure, while true, can be amplified. The purpose of the keel, a hefty built-up structure of trapezoidal cross section, was to structurally bridge the long gap on the fuselage underside caused by the large tandem bomb bays that were as big as a B-17's. The keel extended forward from the front bomb bay opening about six feet to firmly attach it into the forward fuselage structure. It also extended rearward about five feet from the aft bomb bay for the same reason. The four heavy longerons ran the length of the fuselage and were of "T" cross section. Sixty fuselage former rings, or in some cases bulkheads, held the five husky longitudinal members in proper position. An additional seven stringers were affixed to the top of the fuselage before the large, thick covering plates were flush riveted, and in some cases welded, in place. Use of the massive plates cut down on drag-producing joints but required that a relatively new forming technique be used for their fabrication. Called stretch forming, the process had been developed and used in the automobile industry as a high volume production method for forming body panels; however, due to relatively low numerical requirements for any one airplane design, it had been virtually unused in the aircraft industry. Stretch forming is a process by which the work metal, usually sheet steel or sheet aluminum, is stretched over a lubricated wooden form and stressed beyond its yield point to produce the desired contours. Large sheets can be formed into smooth compound curves by means of longitudinal and transverse stretching during which the sheet metal is grasped in jaws that are pulled by powerful hydraulic cylinders. Sometimes the parts are annealed (heat treated) to relieve the residual stresses in the material left by the forming process. In the case of aircraft parts, annealing was required to combat aerodynamic fatigue. The sleek fuselage was built up of four major subassemblies—a nose section, center section, rear section and tail section—all covered with the large stretch-formed plates.

The bombardier's nose enclosure could scarcely be called a "greenhouse" on the B-26; goldfish bowl would be more like it. This one-piece clear plastic molding was the largest ever produced up to that time. No plastics molder wanted any part of it. However, through Magruder's cajoling Rohm and Haas took a shot at it, were successful, and the nose enclosures started moving through the supply pipeline in quantity.

The pilots' compartment was more conventional. It was covered with an aluminum frame and transparent pane canopy. The roof was designed to swing open, parting from the middle in two sections as an escape hatch. The only other glazed-in position was the pointed, streamlined tail gunner's station.

Generally, while the fuselage was built in four sections, these were really major subassemblies that were made up of a total of thirteen smaller subassemblies. The forward section accommodated the pilot, copilot, bombardier, navigator and radio operator. The center section enclosed the two bomb bays, forward and aft. The rear ventral gunner, turret gunner and tail gunner were housed in the aft section.

Aircraft entrance was through a hatch in the nose wheel well leading to the forward crew positions. It could not be used, of course, when the nose wheel was retracted, so there was a bizarre set of emergency instructions for lowering the wheel in flight. Paraphrased it goes something like this: "Maladjustment of screw on nose wheel door operating mechanism arm will prevent nose wheel gear load and fire valve from operation when it is desired to lower landing gear. The main gear will operate normally but the nose wheel will not unlock. It is essential that the crew be familiar with location of this valve as it will be necessary to 'feel' for it when following this procedure . . . [It goes on to tell what fingers to place where, etc. Now comes the bizarre part:] Should it be impossible to locate valve with fingers . . . measure back five inches from

Three-quarter rear and side view of Martin B-26B (Marauder IA) which had a single hand-held flexible .50 in the tail. Marauders now had self-sealing tanks in B model. (USAF)

bottom rear of brake control cover (this is the raised cover extending back from between rudder pedals on pilot's side) and one inch over to the left from the inside edge of the right hand track of pilot's sliding seat. This point will be approximately over the load and fire valve. Carefully gouge a small hole (3 inches to 4 inches square) through the dural floor. Extreme caution must be used as engine control cables and hydraulic lines are under this point . . ." This sure seems a little far-fetched as an emergency procedure!

The nose gear doors were hydraulically operated and were in the down position except when the strut was retracted. Maximum speed for lowering the landing gear was 150 miles per hour. This of course limited the speed at which the nose wheel doors could be opened for bailout.

The aft section of the fuselage was entered through the camera hatch. A ladder to reach that door was stored on the rear starboard wall of the fuselage. There were also emergency escape hatches including the ones through the top of the pilot and copilot's enclosure, out the navigator's observation hatch, through the bomb bay doors and, of course, out the camera hatch. As a convenience, one key fit both the front and aft hatches and all doors except those on the bomb bay could be opened either from the inside or outside. The other openings into the fuselage were the bomb bay doors, which were of two different types. The front bay doors folded upon themselves as they opened. This was arranged because of the very wide opening created by the large-diameter fuselage at this station. The folding method gave adequate ground clearance for large bombs to be carted under the aircraft for hoisting up to the bomb racks. The fuselage diameter of the rear bay was enough smaller that the rear doors simply opened from the center-line down and outward, the more conventional method.

The wing of the prototype B-26 was small for the fuselage size, spanning 65 feet with an area of 602 square feet—eight square feet less than the much

Top view of B model Marauder in which the spinners were deleted. Round circle on fuselage top near wing leading edge is the cover for the retractable navigator's bubble. This photo clearly shows the short wingspan/high wing loading/hot handling characteristics of the B-26. (USAF)

smaller bodied B-25. The prototype and initial production models, because of the resulting high wing loading, required a somewhat higher degree of piloting skill than was available from the new cadets-turned-AAF-pilots. Therefore changes were made in the aircraft wing and tail, primarily a new span of 71 feet and an increased area to 659 square feet. Height was also increased from 19 feet 10 inches to 21 feet 6 inches, to provide a larger vertical tail surface to offset the increased wing span in the yaw mode.

The wing structure was made up of twenty-three premanufactured subassemblies including the center section. Each wing consisted of three leading edge assemblies, a forward box spar section of tension field construction, a rear box span section of the same construction, two flaps, a nacelle fairing, aileron, trim tab, and a removable wing tip. The outer skins were thick, as on the fuselage, and as a strong and novel approach, corrugated aluminum was used instead of the usual stringers for spanwise stiffness. The riveted structure was as sturdy as a bridge and no Marauder ever suffered structural failure no matter what air loads were exerted on it. This sturdiness also helped greatly in absorbing battle damage. For ease of maintenance, another feature (usually found on jet fighters today) was a continuous hinge between the leading edge members and the box spar primary structure. The airfoil shape was dictated by about twenty ribs in each wing starting with a root section NACA 0017 64 and blending outward to a tip section NACA 0010 64. These were medium-lift symmetrical sections primarily designed for speed. The shape of the fuselage in profile was derived from the airfoil sections. The wing was set in the fuselage center section at a 3°30' angle of incidence—the same as the B-25. Dihedral was 1°19' 54". The ailerons were doped fabric covering over metal frames with all-metal trim tabs. Total aileron area was 34.4 square feet and travel was 20 degrees up and 15 degrees down.

An inboard and outboard slotted (after the B-26B) flap was located on either side of the nacelle on each wing. They were hydraulically operated by means of a lever on the pilot's control pedestal. Having an area of 71.82 square feet, they were lowered on takeoff varying amounts dependent upon conditions of field surface, gross weight and need for a quick altitude gain because of conditions such as trees or a hill. Generally speaking, a ½ down position was best for fastest climb while the ¼ down position got the aircraft off the ground in the shortest possible run. By the time climbing airspeed reached 150 miles per hour, the flaps should have been retracted. A landing with ¾ flaps and with power did nicely. As the pilot became more proficient, less flap and less power were usually used. In case of combat damage or hydraulic failure the flaps could still be lowered with an emergency mechanical hand crank located on the forward bomb bay's rear bulkhead.

The horizontal tail surfaces spanned 28 feet and were set at an angle of incidence of −½ degree and 8 degrees dihedral. The maximum chord was 8 feet 2 inches with an elevator area of 69.3 square feet and stabilizer area of 160.2 square feet. Elevator travel was 20 degrees up and 12 degrees down. Elevator tabs were set in the trailing edge and had an area of 3.5 square feet with 15 degrees up and 35 degrees down travel. These tabs, though small, exerted forces on the control surfaces in such a manner that the pilot did not manually have to offset loads induced by the surfaces while keeping the airplane trimmed.

The vertical tail had a fin area of 49.2 square feet and a rudder area of 33.1 square feet. Rudder travel was 25 degrees right or left. It too had a trim tab with an area of 2.1 square feet and travel 15 degrees right or left.

Construction of the fin and stabilizer were of two tension field type beams and sheet metal ribs built into a box cantilever stressed skin structure. The elevators and rudder were metal frames and ribs with doped fabric covering. All tail control surfaces were statically and aerodynamically balanced.

The final airframe components were engine nacelles consisting of four major subassemblies of stressed skin construction plus two massive tubular frame structures acting as engine and landing gear mounts. The main landing gear doors were open when the gear was down and after retraction were closed hydraulically on early models, mechanically on later models.

To power this sleek but somewhat hefty airframe required the most powerful engines available at that time, the R-2800-5 Pratt & Whitney Twin Wasp. This was a giant 18-cylinder 2,300-pound, turbo-supercharged mechanical marvel. Pratt & Whitney, like Ferrari, believed in small bores and short strokes. The result was a smooth running engine with less vibration. The engine had a gear reduction ratio of 2:1, i.e., the propeller shaft ran at only half the engine crankshaft rpm. This was accomplished by means of a set of reduction gears located in the nose of the engine crankcase. Each engine delivered 1,850 horsepower at 2600 rpm during takeoff. (Pushed all out, they could produce 2,000 horsepower at 2700 rpm.) Throttled back to 2400 rpm at 13,000 feet they each still produced 1,450 horsepower.

The propellers used to harness this tremendous horsepower were 13 foot 6 inch diameter Curtiss electrics with four C-542-S-814CC2 blades. One propeller alone weighed in at 594 pounds. The engine throttle and propeller controls were located on the pilot's pedestal in the center of the cockpit.

Four self-sealing fuel tanks were located in the wings. Two were inboard from the engine nacelles and were the main tanks from which the engines were

The B-26B moved along at a maximum speed of 311 miles per hour at 14,500 feet. Some were fitted with an external rack that carried a 2,000-pound torpedo. (USAF)

This B-26B model was fitted with enlarged carburetor air intakes to accommodate the oversized air filter used in desert combat. (USAF)

fed, and the two outboard of the nacelles were the auxiliary tanks. The wing tanks and two drop tanks located in the forward bomb bay were emptied by means of electrically driven booster and transfer pumps into the main tanks which then supplied the engines. This process had to be carefully monitored, as once the main tanks were full, continued pumping would burst them or overflow them into the wings and bomb bay. Another problem connected with the fuel tanks was that there was no baffle provision for side slip conditions. Such a maneuver generally caused the engine on the low wing to quit. At low speed and low altitude this was the last thing in the world you wanted to happen in a B-26. Fuel transfer control switches were located on the pilot's pedestal. The normal fuel load was 962 gallons; however, for ferrying it went up to 1,212 gallons. That sounds like a lot but at takeoff and emergency maximum speed the engines were burning 266 gallons per hour. Even with all stops pulled for maximum economy, consumption was still 116 gallons per hour. Normal continuous maximum speed, such as used on a mission, used up 169 gallons per hour.

Lubricating oil for the engines was contained in oval 41.25 gallon tanks located at the rear of each engine mount. The oil cooler radiators were located beneath the tanks in the lower portion of the nacelles. Cooling air for the radiators was admitted through scoops located in the lower leading edge of each cowl. Hydraulic exit flaps for the air controlled the amount of airflow, thus maintaining the correct oil temperature. Position of these flaps was regulated by a control on the pilot's pedestal. Ideally, temperatures were held to about 150°F; however, for maximum operation such as all-out climb or speed, it could go to 212°F without harming anything. Oil consumption was 60 quarts an hour for the two engines at maximum performance but dropped to 32 quarts an hour at economical cruise speed. And you thought the old Studebaker burned oil!

Atop each of the engine cowlings were two big carburetor air scoops to direct air through large rectangular filters before it entered the carburetors. These were needed for use on unimproved runways. Engine cylinder head temperatures were controlled by hydraulically operated cowl flaps much as those on the previously covered B-25J. Cylinder head temperatures ran at about 500°F.

Close-up view of the 18-cylinder Pratt & Whitney R-2800-5 engine that delivered 1,850 horsepower at 2600 rpm. The four-blade Curtiss electric full-feathering controllable-pitch propeller can also be seen. Two large scoops on cowl top are for carburetor air; lower scoop is for oil cooler. (Mendenhall)

Nose wheel and strut detail of B-26. The strut swiveled 90 degrees on retraction, placing the nose wheel flat against the pilot compartment floor. (Mendenhall)

Nothing was ever simple on the B-26 and the hydraulically activated tricycle landing gear was no exception. The nose gear swung straight back but the wheel and strut, by means of a bevel gear sector, rotated 90 degrees to nestle in the fuselage, flat against the pilots' compartment floor. The nose wheel mounted a smooth-contour 8-ply 35-inch-diameter tire. The strut was of Martin design and dampened shock loads hydraulically with a total travel of 11 inches. It could swing 58½ degrees either side of the fuselage centerline. At this angle the aircraft could be swung about either main gear locked up by braking, although this procedure was frowned on. Aside from the emergency procedure mentioned earlier the gear could also be hand cranked down from the bombardier's compartment.

The main gear retracted forward into the engine nacelles—but only after going through a wild contortion where the top end of the strut went down and rearward as the end with the wheel on it swung upward into the nacelle. Activation was by means of a small hydraulic cylinder. Travel of the oleo strut was 9-⅞ inches and the tires were 47 inches in diameter with a rolling radius when loaded of 20 inches. The three shock struts that made up the landing gear were all locked in place (up or down) with spring loading and unlocked with hydraulic pressure. A hand pump was provided to build hydraulic pressure for lowering the gear in emergencies—if you still had hydraulic fluid.

Braking was accomplished hydraulically and applied by means of pedals mounted atop the rudder pedals, but only on the pilot's side. With the hydraulics out a 1,000 psi air bottle supplied enough braking force to accomplish one landing. The emergency air brake handle was located over the pilot's head. By opening and closing the handle some control could be obtained over the pneumatic system though care had to be taken not to lock up the wheels completely. After coming to a stop the system had to be depressurized before taxiing. Also in the braking system parking brakes and control surface locks were available and controlled by levers on the pilot's pedestal.

A walk-through of the fuselage, starting with the bombardier's clear enclosure, showed it to be a fairly roomy workplace. The bombardier was equipped with a .50-caliber flexible machine gun mounted nearly in the center of the nose with an ammunition box and spent shell bag to his right. Also, on the right compartment floor on some models was another fixed .50-caliber gun to be fired by the pilot. In the center of the floor was a cushioned seat, positioned to comfortably operate the bombsight mounted over the optically flat portion of the transparent nose. The left compartment wall contained a large map case, an electric light, oxygen regulator and radio jack boxes. On the floor to the left of the seat was a small pedestal with a lever. That was the bomb control quadrant. On the right compartment wall was the instrument panel with altimeter, compass, airspeed indicator, outside air temperature, sweep hand clock, etc. Also on this wall was the intervalometer, electric bomb selector handle, and bomb bay door controls. In addition there were air ducts for hot and cold air and defroster tubes. A removable bombsight box with lock and key was provided for the bombardier's use. Entry to the compartment was through a door in the rear bulkhead separating the pilots' compartment from the bombardier's. With two voluminous bomb bays the bombardier had a wide variety of loads to choose from. Peyton Magruder had gone to Wright Field and, with the aid of a tape measure, jotted down the dimensions of a B-17's bomb bay. This size then defined the front bay; to this he added a second smaller bay right behind it. The bombs were shackled to two heavy structural vertical uprights that went from the keel to the bomb bay's ceiling. These uprights were canted outward in the shape of a tall thin "V" so that the upper bombs would clear the rack as they fell out of the bay. A hand-crank bomb hoist was located between the two racks. The forward bay would hold any of the following combinations: twenty 100-pound bombs, eight 300-pound bombs, six 600-pound bombs, four 1,100-pound bombs or two 2,000-pound bombs. The smaller rear bay could hold ten 100-pounders, six 300-pounders or two 600-pound bombs. The aft bay, by the time the "C" model B-26 came about, was eliminated.

Immediately behind the bombardier's compartment was, of course, the flight deck. The pilot's seat was well armored with ⅜-inch plate protecting his head and a ¼-inch plate on the seat back. The copilot's seat had only ¼-inch armor plate on the seat back. The instrumentation was, as in most twin-engine military aircraft, somewhat exotic to the uninitiated. However, broken into basics, the instrument panel consisted of a flight attitude group, a navigation group and an engine monitoring group, along with a few military innovations such as bomb bay door position indicators. Though not as sophisticated, particularly in the radio equipment, it was generally the same instrumentation found in your twin Piper, Beechcraft or Cessna of today. One big difference was that the instrument panel did not extend across the whole cockpit—it was mainly for the pilot's use. The copilot had a yoke and pedals (which could be folded out of the way) but had to crane his neck to the pilot's side to view the instruments. As a matter of fact, some B-26's were built like the British bombers, even the four-engine heavies, without a copilot's position at all! This idea was soon canned by the aircrews as there *was* a certain comfort with two pilots up front instead of one. (The British were short on pilots but our massive training program created a plentiful supply.)

The cockpit sides contained jack boxes, headset plug-ins, oxygen regulators, map cases, side window latches, etc.

Dual side by side wheel-type control columns and rudder pedals provided the usual flight controls. Adjustment for leg length was accomplished by moving the seats forward and rearward on tracks. The tracks for the copilot's seat were angled inward from the right cockpit wall so that as the copilot moved his seat forward he shifted more toward the center of the aircraft, thus allowing him a better view of the instrument panel. On the rear bulkhead of the cockpit was a thermos water bottle and cup holder, a fire extinguisher and two oxygen bottles. The center of the bulkhead had a doorway leading to the navigator's compartment. A split draw curtain covered this doorway to help cut down on the cold drafts that flowed through the plane at altitude.

The navigator and radio operator sat side by side in the compartment between the cockpit and the forward bomb bay, the radio operator on the left and navigator on the right. An inflatable life raft was stowed in the ceiling of this compartment. The navigator was equipped with the usual map table, map cas-

Pilot's instrument panel of B-26 showing control column, rudder pedals and center pilot's console. (USAF)

Flight Instruments

- Airspeed indicator
- Alternator
- Turn and bank indicator
- Rate of climb indicator
- Directional gyro
- Gyro horizon
- Pilot director
- Clock
- Radio compass with marker beacon light
- Outside air temperature
- Magnetic compass

Engine Instruments

- Manifold pressure gauge
- Tachometer
- Fuel pressure
- Oil pressure
- Oil temperature
- Cylinder temperature
- Fuel flow meter
- Carburetor air temp.
- Fuel pressure lights
- Liquidometer (one only)

Systems Instruments

- Flap and landing gear position indicator
- Engine synchronization
- Air suction
- Hydraulic pressure
- Fuel mixture
- Landing gear lock pin lights, left side; nose, right side
- Instrument lights knob
- Air pressure
- Deicer lights—on or off

Flight and Systems Controls

- Blowers and guard
- Hydraulic pressure gauge
- Wing flaps, control lever
- Right and left cowl flaps, control levers
- Parking brake lock lever
- Throttle lock
- Right and left throttles
- Elevator controls
- Circuit breaker, propeller
- Starter energizing and mesh switch
- Left and right landing light switches
- Master switch ignition
- Signal light switch, interior
- Left and right engine primer switches
- Alarm bell switch
- Formation light rheostat
- Left and right fuel booster switch
- Wing position light switch
- Left-hand propeller toggle switch
- U/V instrument light switch
- Inverter switch
- Left and right battery switches
- Left and right oil dilution switch
- Right-hand propeller toggle switch
- Fluorescent light
- Circuit breaker, propeller
- Left and right propeller feathering switches
- Mixture control levers
- Propeller governor control
- Tail position lights switch
- Landing gear emergency control lever
- Propeller governor control lock
- Landing gear lever
- Identification light switch box
- Pedestal light switch
- Carburetor air control levers
- Oil cooler shutter controls levers
- Compass light switch
- Pitot heater switch
- Left and right magneto switches

A more detailed view of the instrument panel. The panel did not extend completely across the cockpit, leaving the right side open for the bombardier's entryway into that station. (USAF)

es, radio compass, jack box and oxygen regulator. An astrodome was mounted ahead of the life raft in the compartment ceiling so that celestial navigation fixes could be taken with the sextant during long overwater flights. Next to the navigator's table, the radio rack contained an array of communications equipment such as the radio compass, loop antenna controls, command radio receiver and transmitter, marker beacon receiving equipment, antenna reel, interphone equipment and a liaison transmitter.

A hatch on the rear wall of this compartment allowed entry into the front bomb bay. Moving through the front and rear bomb bay, a hatch led to the aft section of the aircraft. Each bomb bay had a relief tube station and there was a third one available for the tail gunner in his area. (In addition, a chemical toilet was provided beneath the dorsal turret.) Ammunition boxes for the dorsal turret were sometimes stored in the rear bay after its use as a bomb carrier was abandoned.

The power turret was of Martin design, a Model 250-CE, and mounted two .50-caliber machine guns. The main member of the turret was a large aluminum alloy casting to which was fastened the brackets and mounts for the gun carriers, gun sight, seat, armor plate apron, ammunition boxes, plexiglass enclosure and power drives. The turret, rather than being pedestal mounted as was the B-25's Bendix turret, was attached to the top of the fuselage. It was suspended on twelve pairs of ball bearing rollers. A stationary ring gear around the turret was fixed to the fuselage mount. An electric motor on the turret drove a small pinion gear that meshed with the ring gear. In this manner the turret could be revolved in azimuth. A similar arrangement, except in the vertical plane and attached to the gun carriage, elevated and lowered the guns. Because of weight and size limitations these drive motors were necessarily small, which meant there were operating limitations to prevent burning them out. The turret was designed for the following duty cycles:

1. Successive 30 minute cycles consisting of a period of 10 minutes continuous azimuth operation at full speed followed by 20 minutes of power off.
2. A period of 4 minutes with the turret rotating back and forth through 180 degrees with a maximum acceleration of 60 degrees per second and a maximum speed of 60 degrees per second.
3. A period of 4 minutes with the guns elevating and depressing through 90 degrees with a maximum acceleration and deceleration of 40 degrees per second and a maximum speed of 40 degrees per second.
4. A period of 1 minute every 30 minutes with the guns stalled or operating at speeds not in excess of 1 degree per second against maximum windage loads.

What all the above means is that during a long-running fire fight with enemy fighters, the gunner had to be a little careful not to put his turret out of operation due to burned-out drive motors.

Another feature of the turret was automatic azimuth fire interrupters so that the gunner didn't shoot his own tail off in the heat of combat. He still had to worry about the propellers, though, as that portion of the firing arc was not automatically interrupted. Operation of the turret—once the gunner was seated and strapped in—was initiated by reaching under the front edge of the seat and closing the two main power switches. The gunner then closed the gun safety switch on the right-hand side just above the seat. With both hands placed on the firing grips his left hand closed the dead man switch—the turret would not operate without this being closed. To elevate the guns he pushed the grips forward and to depress the guns he pulled the grips back. To rotate the turret he twisted the grips about their vertical axis and the turret rotated in the direction the grips were turned. The speed the turret turned depended on how far the grips were twisted. All motion of the turret stopped by allowing the grips to return to their normal spring-centered positions or by releasing the dead man switch. The actual firing of the guns was accomplished by depressing the trigger switch on the right-hand grip. A microphone switch was on the left-hand grip.

The Marauder from head-on looked massive but streamlined. Package guns, dorsal turret and large carburetor air scoops mar an otherwise perfect streamlining job. (USAF)

A study in aero sleekness and beauty, this B-26B-40 skims swiftly along just above the broken cloud deck. (USAF)

Compare this top view with the earlier one in this chapter. The longer 71-foot-span wing is clearly seen to be increased in size over the earlier 65-foot span. Instead of helping landing speed reduction, the Air Force just loaded the ship up more and the wing loading and landing speed remained about the same. (USAF)

This B-26C was the same as the B-26B-10-NA except it was built in Omaha. Note both the absence of spinners on the aircraft's propellers and the troop glider in the background. (USAF)

Although this Marauder is unpainted, the fabric-covered control surfaces seem to be painted olive drab instead of the usual aluminum to match the aircraft. (USAF)

As this B-26B takes off with partial flaps it reveals the wavy color line between the gray fuselage bottom and olive drab sides and top. The color line of the nacelles was straight. (USAF)

Against a bright blue sky spotted with cumulus clouds the optically flat bombsight window shows up well on the Plexiglas nose. (USAF)

An additional two .50-caliber guns were mounted just aft of the turret where they served as flexible waist guns. Ammunition for the waist guns was mounted on the ceiling above each gun. Directly behind these guns was the camera mount located over a hatch in the middle of the floor.

Aft of the camera area was the tail gunner's position. The initial B-26 had a single hand-held .50-caliber Browning with 400 rounds, but by the time the "B" models arrived two Brownings were installed in a stepped-down tail position with 1,500 rounds supplied to each gun from ammunition boxes stored in the aft bomb bay. The ammunition was carried along roller tracks on either side of the fuselage and the linked shells sat upright on their butt ends. The tail gunner was protected with thick bullet resistant glass and an armor plate bulkhead. He sat on a small bicycle seat.

Another item of interest on the B-26 was the parachute stowage. A quick-attachable-type parachute was stowed at each of the following positions: bombardier, navigator, radio operator, plus four in the aft compartment. Seat or back chutes could also be used by the pilot, copilot, navigator or radio operator by removing one of their seat cushions.

Due to the tremendous noise involved during combat, cotton containers were provided at each crew station for use as ear plugs.

For winter or cold weather operations, wing and tail deicer boots were fitted; their operation was similar to those described on the B-25. For propeller deicing, standard slinger rings were provided with an electric-driven pump and suitable tubing conveying the anti-ice mixture from a five-gallon tank on each engine. Care had to be taken to not let too much ice build up on the props before using this device to avoid hurling large chunks of ice off the blades.

Heat for the aircraft was supplied to all crew stations by means of exhaust manifold heaters. After combat this system was generally shut down due to the possibility that bullet holes might have perforated the exhaust manifold, which would allow carbon monoxide into the heating system.

One last item for cold weather operation was a zippered curtain in the aft compartment forward of the tail gunner's entrance to maintain better heating and ventilation for him.

All in all, the B-26 was a thoroughly engineered, sophisticated flying machine that, like the B-25, heralded the construction and techniques used on jets in more modern days.

With the starboard engine out and electric propeller feathered, the Marauder still flies! This is contrary to what many people said of it in its early days. Note the great amount of left rudder the pilot had to feed in to accomplish this, however. (USAF)

6

THE MARAUDER MISSIONS

THE MARAUDER CAME OUT OF the ETO at the end of the war smelling like a rose, but it wasn't always that way. To begin with, its career got off to a bad start. The initial Bomb Group to be outfitted with it was the 22nd at Langley Field, Virginia. By June of 1941 the first problem arose and production was stopped. Of 66 planes built, 44 remained stored awaiting repair and rebuilding of the nose wheel strut which now seemed too weak for operation. The struts were strengthened but the real cause of the failure incidents was traced to the Army itself. Martin, unable to provide armament for the first batch of aircraft, had balanced the early B-26's with tools and spare parts in its place. The Army, neat as it always was, removed the ballast and as a result the ships were nose heavy—so much so that the nose gear collapsed when slammed down on a 130 mile per hour B-26 landing. This situation was straightened out and once more production commenced.

Pearl Harbor day came and went, the nation's defenses on that island stronghold almost totally ruined. The next day the 22nd Bomb Group headed for Muroc, California, to begin submarine patrol duty, a job it kept until the end of January 1942. The Group's 57 planes were then moved by sea to Hawaii, where they were fitted with bomb bay tanks, and flown on to Brisbane, Australia, completing that journey by February 25.

After more training and getting the group acclimated to their new base they flew their first combat mission against the large Japanese supply depot at Rabaul. The trip was made in stages from Townsville, Australia, over the Coral Sea for about 700 miles to Port Moresby, New Guinea. Here the B-26's were refueled for another 550-mile leg to the target at Rabaul on the northern tip of New Britain. Two thousand pounds of bombs (either four 500-pounders or twenty 100-pounders) plus a 250-gallon bomb bay tank were taken by each plane on the trip. The Marauder raids covered 16 missions and 80 sorties to Rabaul before they were finished on May 24.

Also raided during the same period were the Lae enemy bases which were only about 250 miles away on New Guinea itself. This route, though short, was a tough one as it was over the Owen Stanley Mountains and the thick New Guinea jungles. On these missions the Marauders were often bounced by the Japanese Zeros, but only three were lost in the 84 sorties flown against Lae, ending on July 4, 1942.

By this time the battle of Midway was shaping up. Four Marauders had been split off from the 22nd on the way to Australia and remained in Hawaii for (of all things for an Army medium bomber) torpedo training! A rack was perfected to mount a standard 21-inch-diameter torpedo under the aircraft's keel member. In addition to the rack, a goodly portion of the 2,000-pound load was carried by cables that reached up into the internal bomb rack structure. With a do-or-die massive naval battle shaping up, the four twin-engined torpedo bombers were hustled over to Midway to await the chance to sink an enemy warship. It didn't take long. On June 4, 1942, the four were in the air pi-

This shark-nosed monster, 42-107582 (a B-26C-45), was with the 323rd Bomb Group in England. Upper surfaces of fuselage were sprayed medium green. And remember the Quonset hut—there's one in the left background. (USAF)

With 111 missions emblazoned on its flank, "Ish-Tak-Ha-Ba" of the 9th Air Force sets ready to perform yet another. Note Indian brave's head on nose wheel. (USAF)

Dawn at an English field breaks over two 9th Air Force B-26's. They have long since been gassed up and the armament people have completed their jobs for the upcoming day's mission over Germany. (USAF)

Rollin! "McCarty's Party" of the 391st Bomb Group moves out on March 8, 1944, in England. (USAF)

A B-26B-2 called "Hell Cat" had already knocked off 50 missions for the 17th Bomb Group when this photo was taken. (USAF)

A six-plane takeoff from a desert strip in North Africa can mean only trouble for Rommel's Afrika Korps. (USAF)

North Africa and this B-26, with nose wheel retracted and the rest of the gear on the way up, heads off for its daily mission. (USAF)

loted by two lieutenants from the 22nd Bomb Group, Mayes and Muri, and two other pilots, Captain Collins and Lieutenant Watson of the 38th Bomb Group. At 800 feet they began their runs at the enemy carriers. The protective Zeros drove them down to a scant ten feet above the water. No hits were scored and the angry Zeros destroyed two of the four would-be torpedo bombers. Collins and Muri, carrying heavy battle damage, limped back to Midway.

North of Midway, the Japanese had pulled a diversionary raid on Alaska's Dutch Harbor the same day. B-26's, twenty-four in all, were waiting for them. The aircraft were assigned to the 28th, 73rd and 77th Bomb Groups. Carrying torpedos, like their Midway brethren, they took to the air in search of the enemy carrier that had launched the attack against them, the *Ryuyo.* They finally spotted it but no hits were scored. That was about it for the Alaskan B-26's. No one else ever came up to that portion of the world to fight. The Marauders, by now thoroughly evaluated against the nimble Mitchell, was losing popularity fast in the Southwest Pacific. The gentle B-25 could take off more

"France Flies Again" was the motto of the French Air Force's B-26 Marauder Group that operated in the Mediterranean theater. (USAF)

A B-26 dispersal area at Chipping Ongar, Essex, England, in 1943 shows five of the aircraft parked on their hardstands. (USAF)

quickly, land slower and fly further, and it had one other advantage: greater propeller to ground clearance. Most jungle fields with compacted soil and steel matts were rough compared to the smooth runways of the United States and Europe. No two ways about it, for the rough-and-ready low-altitude combat of the islands the B-25 was the superior aircraft—less sophisticated and easier to maintain without a full-blown maintenance facility. Mitchells replaced Marauders in the 39th Group's inventory in September 1942, followed with a changeover by the 22nd Group in early 1943. That mainly ended the Pacific career of the Martin B-26 but it was soon to be used heavily in the ETO.

The color scheme of these initial Pacific Marauders was the usual olive drab topside (sometimes splotched with green) and pale gray underside; that is, except for the "Great Silver Fleet." That name derived from the slogan embla-

A B-26B-1 (41-17834) and two B-26B-40's (42-43269 and 42-43267) show the difference in tail gunner positions between the two models. (USAF)

The 9th Air Force heads for Germany en masse with its B-26 Marauders. Twenty-eight aircraft appear in this photo taken over the English Channel. (USAF)

zoned on the sides of Eastern Air Line's prewar fleet of shiny aluminum DC-3's. It was now adopted by the 19th Bomb Squadron of the 22nd Bomb Group. They had been equipped with reconditioned B-26's and part of this reworking required that the olive drab paint be removed for inspection purposes. It was not replaced. A winged circular insignia was designed for the vertical tail and it was labeled "Silver Fleet." The squadron operated these silvery misfits proudly until January 9, 1944, when they were converted to four-engine B-24's.

Nineteen forty-two, and Egypt was in trouble. Rommel and the mighty armored Afrika Korps were steadily pushing the British, under Montgomery, toward the Nile. In an effort to help turn the tide, a lend-lease group of 52 Marauders went to the Middle East via the northern route—Gander, Greenland, Scotland and England—during September 1942. Three of the B-26's stayed in England for tests while the rest continued to Egypt, arriving there in November and December. The first mission was a recon sortie involving only one Marauder but soon they were all bombing the Tunisian ports and cities that were

Enemy installations are bombed by the 12th Air Force's B-26G-5's somewhere over southern France. (USAF)

The "Silver Streaks" with war paint removed for greater speed. These Marauders belonged to the 397th Bomb Group stationed at Rivenhall. The tail strip was yellow, edged in black. (USAF)

A truly beautiful photograph of Marauders in flight. The unit is returning from a mission along the French-German border in support of the U.S. Third Army. (USAF)

This photo of a B-26G-5-MA (43-34396) shows well the engine thrust line and the wing incidence angled upward 3.5 degrees. (USAF)

Wing skin wrinkling due to rough air flexing shows up on the Marauder of the 320th Bomb Group. Fin numbers and stripe under stabilizer were yellow, edged with black. (USAF)

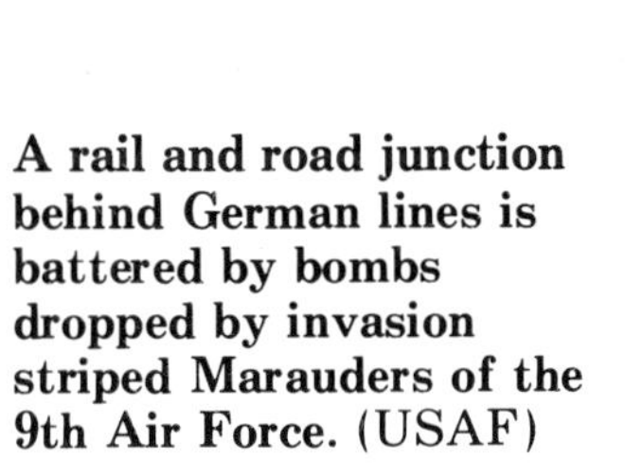

A rail and road junction behind German lines is battered by bombs dropped by invasion striped Marauders of the 9th Air Force. (USAF)

A 9th Air Force B-26 heads for home after a successful attack on objectives along the French coast. (USAF)

This remarkable photo was made at the instant flame billowed out from the Marauder's ruptured fuel tanks. The flames nearly envelope a trailing aircraft. Flak did the damage as the formation hit a target in the Pas de Calais area. (USAF)

No story or book about the Marauder ever fails to use this dramatic photo—and justly so. The R-2800 engine, even detached from the aircraft, still runs from the fuel left in the carburetor bowl. A flak hit from a German 88 at Toulon Harbor, southern France did the job during a raid on the harbor's coastal defense guns. (USAF)

With bomb bay doors open, and lower longitudinal keel showing, a stick of demolition bombs heads for the target. YU-Y is from 386th Bomb Group. (George F. Wegman)

used by the Germans to supply their armies. (On December 16 there was an unfortunate accident. A group of Spitfires, not recognizing a B-26 due to its unfamiliar lines, shot the B-26 down.) Soon it became obvious that sand filters would be needed for desert operations and this is when the tops of the engine nacelles were fitted with large air scoops to house such filters. Three B-26 Bomb Groups, all part of the 12th Air Force under the command of Jimmy Doolittle, moved into North Africa shortly thereafter. In November 1942 the 319th arrived, the 17th arrived a month later, and in April 1943 the 320th arrived.

The 319th, flying the short-winged B-26B, had flown to England in September of 1942 and on to Algeria in November. The 17th flew its first mission on December 30, bombing Tunis. They even got in a few shots at the giant Me 323 troop transports.

Before that period, however, Marauders were coming off the assembly lines and going to training centers. At Barksdale and MacDill they found themselves again in trouble. Due to many divergent reasons the new aircraft was rapidly earning a bad name for itself. The inexperienced training groups were putting pilots with only a couple hundred hours into this hot, fast aircraft. "The new pilots were afraid of the B-26 and we had one accident after another," recalled Hap Arnold in his memoirs. Seemingly, all that was necessary was for one engine to go sour on a B-26 in flight and it would crash. "A plane a day in Tampa Bay" became the slogan for the student pilots. Names for the plane (it started out "Martian," then was changed to Marauder) were less than complimentary. "Baltimore Whore," the "Widow Maker" and the "Flying Prostitute" were much more common names for the aircraft.

At the request of Wright Field Production Engineering all B-26's were grounded April 21, 1941, due to nose gear failure. As mentioned earlier, a similar problem had occurred due to the center of gravity being changed when the ship had no armament. This time it was different—the armament was in place. An investigation turned up that the nose gear parts were improperly heat treated and, as a result of embrittlement, broke on impact. While not the design's fault, General Arnold decided the time had come to re-evaluate the Army's decision to build the bomber. He asked his old friend Jimmy Doolittle to investigate, telling him to take a B-26, fly it under any and all conditions, then go to MacDill and take temporary command of the training groups. Doolittle evaluated the B-26 and decided it was a good aircraft. The level of piloting skill was a somewhat higher requirement than for, say, a B-18 or B-25, but the plane itself was OK. His first recommendation was that they continue building it.

Arriving at MacDill, he put on a flight demonstration with the aircraft that left student pilots breathless, including single-engine landings and takeoffs, and dead-stick landings. The student pilots, who had been selected from the top third of their Cadet classes for flying ability, regained their confidence, caught the spirit and soon they too were flying and landing the B-26 with one engine. That took care of the people side of the problem. Doolittle summed it up later in an interview: "The B-26 Marauder was an unforgiving airplane and it was killing pilots because it never gave them a chance to make mistakes." It can be added, however, that given the right training to competent pilots, no mistakes were made. As the press said, it *was* a Flying Torpedo and no slipshod, ill-trained pilots need apply.

Mechanical problems, however, continued to pop up, this time in a series of crashes that seemed connected with propeller overspeeding. The Curtiss electric propellers were sophisticated pieces of machinery and as such needed a shakedown period to gain field experience as well as train mechanics to keep them performing well. The failure mode in each crash was the same. The B-26 would race down the runway, Pratt & Whitneys bellowing, rotate, break ground and stall back out of control, destroying almost all evidence of what caused the accident. A propeller going into flat pitch at the wrong time could be the cause of this type of failure but inspection of the propellers turned up no unusual reason for the malfunction. Finally, Wright Field engineers had a hunch. The prop hub pitch setting mechanism was driven directly off the batteries. If the new and green mechanics when checking out the power turret ran the batteries down, the prop overspeed condition could happen. They followed up their hunch and found themselves to be 100 percent correct. Battery carts were developed to provide current for servicing and the propeller circuit was changed to run directly off the generator, with the batteries continuing to be used only for backup power. End of problem—for awhile.

Next came sudden engine failures. An engine would be running great, then just quit for no apparent reason. With a bit of luck a failure finally occurred during an engine ground run-up. This time the problem was determined to be a diaphragm in the fuel control system. It was rubber and worked just fine until

Eight 500-pound bombs begin their descent on an enemy military installation in northern France on May 9, 1944. With flak bursts surrounding it, it's time for the Marauder to get out of there! (USAF)

A B-26B-25-MA and B-26C-5-MO of the 449th Bomb Squadron speed across France heading for home base in England (USAF)

A low-level dust-up is enjoyed by the 26 crews over the home field after return from a successful mission. While against all rules and regs this sort of thing let the steam off after lengthy combat operations. (George F. Wegman)

"Bunny's Honey" with 27 missions under its belt heads for the upper reaches en route to the target area. (George F. Wegman)

A stick of twenty-six 100-pound bombs are salvoed on a Nazi installation in France. Aircraft is a B-26B-50 of the 387th Bomb Group. (USAF)

A B-26B-40, N42-43459 flies along above the clouds on a training flight.

additives were put in the gasoline to make the new high-performance aromatic fuels the military wanted. The rubber couldn't take the chemicals and eventually rotted away. With high manifold pressures the weakened diaphragms would rupture and the fuel flow would stop. All B-26's were again grounded and about 80 percent of the membranes were found ready to let go. A change to a diaphragm material that was unaffected by the aromatics stopped the problem. The B-26's were flying again.

One other problem along this line was the chronic leaks the hydraulic system was subject to. Again a chemical incompatibility between the system's O-ring seals and the new nonflammable hydraulic fluids, required by the Army,

Two blown tires as a result of this B-26G-1's hot landing brings out the fire truck and a crew of mechanics. (USAF)

They said "it couldn't be done" but with a skillful, experienced pilot a B-26 could return to base on one engine. "The Hearse" comes in over the numbers as proof of that. (USAF)

was the culprit. The seals were fine with the traditional Lockheed hydraulic fluid but worthless with the new stuff. Here again a change in rubber compounding solved the problem.

Finally, with the major mechanical problems solved, with pilots now becoming experienced in handling the ship and mechanics now well trained for their jobs, the Marauder was ready for the war even though two or three times the project had been nearly scrapped and production stopped.

In the ETO the 8th Air Force was initially the front running combat unit for the United States. Four bomb groups using B-26's were assigned to it, the 322nd in May 1943, 323rd in July 1943 and the 386th and 387th in August 1943. The first of October, 1943, the 9th Air Force was formed and these above four bomb groups were turned over to it. The 9th then took over three more B-26 groups: the 391st in January 1944, the 394th in March 1944 and the 397th in April 1944. So in April 1944 there were eight B-26 bomb groups in the ETO under the command of the 9th Air Force.

As a matter of interest, the table of organization for an air force runs theoretically like this: An air force is made up of three wings with six bomb groups in each wing. There are four squadrons to a bomb group. The squadron consists of twenty bombers—eight on the ground for maintenance spares and

The "Jolly Roger" with many missions emblazoned on its flanks until the day of a belly landing and then the scrap heap. The assembled personnel look as if they were attending a funeral. (George F. Wegman)

four flights of three aircraft in the air. Therefore, a bomb group has eighty aircraft and the eight B-26 bomb groups equaled about 640 aircraft plus additional spares at replacement holding centers. The Marauders were numerically ready to fly against Fortress Europe—well before D-day.

The first combat for B-26's based in England was on May 14, 1943, a low-level mission against the Velsen generating station at Ijmuiden, Holland. Using 30-minute time delay fuses on their 500-pound bombs the B-26's came in through heavy flak at 100-300 feet. The long-time-delay fuses—used to allow the Dutch workers to escape—backfired, for the Germans defused the bombs before they exploded. Nothing resulted from the mission except for a few holes punched in roofs by falling bombs.

Three days later, on the 17th, the same B-26's were back trying to destroy the power station again. This time they were not so lucky—the Germans were waiting for them. During the trip across the Channel one of the twelve bombers had an electrical failure and returned to its base, Great Saling. The other eleven, as they neared the target, were greeted with extreme flak. Almost immediately the squadron commander was shot down. Another B-26 was hit and collided with a third, the two of them going down entangled in each others' wreckage. The intense ground fire flamed two more as they passed low over Haarlem. On the return flight the other six were knocked down by Me 109's. It could not have been worse—the entire squadron was lost and the name "Widow Maker" once more was heard amongst the B-26 flying people. The Air Force had learned a hard lesson. No more would the B-26 be committed to low altitude missions, at least until control of the enemy in the air and on the ground was obtained.

An early Marauder, probably a B-26B, on the right of the bombsight optical flat, a cover is mounted over the hole that took a fixed .50-caliber machine gun. (USAF)

Invasion stripes on the wings, tents and bomb storage stacks indicate this photo of a B-26 was taken shortly after D-day. (USAF)

After the Ijmuiden debacle the crews were retrained for medium-altitude missions, bombing with the Norden sight from 12,000 to 18,000 feet. By July 17 the Marauders were back in combat bombing German air fields, rail yards and other targets not cost effective enough to use the B-17's and B-24's. With D-day nearly a year away the Marauder continued this type mission with a now super-low loss rate. With the June 6, 1944, invasion they were used all-out in the fighting, their fuselages and wings emblazoned with the 24-inch-wide black and white invasion stripes. They then continued to serve right up to war's end with distinction, being particularly useful in destroying German V-1 rocket launching emplacements. In early 1945 two units did finally change over to the newer Douglas A-26's, the 386th and 391st Bomb Groups.

Additional overseas users of the B-26 included six Free French squadrons based in North Africa, and five South African squadrons used as part of the Allied Desert Air Force, where they continued to fly missions over Italy once the Germans had abandoned the African continent. RAF use was with No. 14 and No. 96 Squadrons.

A B-26B taxies for a training flight. The racks for a 2,000-pound torpedo can be seen on the fuselage belly. (USAF)

A training flight of three Marauders over Southwest Texas brush country. The lead ship is an AT-23B, the other two are B-26C-5's, (USAF)

Back in the United States two Bomb Groups, the 335th and 336th, acted as training units until they were disbanded in May 1944. The Navy also used some B-26's, brightly painted in reds, yellows and blues, as target tugs. Every naval air station seemed to have one or two of the colorful ships.

The B-26's were efficient in war but still remained a handful to fly. Not many made it back to the States. They were blown up en masse in Germany and taken to the smelters rather than bother bringing them home. A few exist today in museums here and there. The Confederate Air Force owns one which it plans to restore and fly some day. While it was working, though, it flew over 110,000 sorties and dropped 150,000 tons of bombs.

A permanent monument to the B-26, a G-10 model, and one of the few complete Marauders left today. It saw action during World War II, flown by the Free French. Today it resides inside the Air Force Museum, restored and spared the ravages of weather and time. (Collect Air Photos)

7

ANOTHER DAY—ANOTHER HEADLINE

DENSE FOG LAY OVER MANHATTAN early Saturday morning July 28, 1945. The heavily populated area looked bleak, skyscrapers pointing their spires out of sight into the somber gray blanket. Above the gray, on top, where the sun made the shroud fleecy white, Colonel William Smith called La Guardia for weather information. He and his copilot along with a hitchhiking sailor were en route from Bedford, Massachusetts, to Newark, New Jersey. LaGuardia radioed back to the B-25, "extremely poor visibility."

Minutes later midtown Manhattan was shaken by an explosion high in the air. The B-25, as it dropped through the fog layer, had collided with the Empire State Building between the 78th and 79th floors and hammered an 18- by 20-foot hole into the building's side. One engine went through seven walls and out the other side of the building to the street below. The other engine hit an elevator shaft, severing cables and sending the car and its woman operator plunging 80 floors to the basement. Part of the fuselage lay broken on the north facade, a wing fell on Madison Avenue, and general debris showered a five block area. Fire from the bomber's ruptured fuel tanks set six floors of the building aflame and out of control for 40 minutes. The avgas also set several fires below on roofs of adjoining buildings. In all, 13 people were killed—three on the bomber. Had the accident occurred on a week day the loss of lives would have been much greater with the normal 15,000 workers in the building plus 3,500 visitors. This day the entire 78th floor was deserted. The accident review board decided Colonel Smith had become disoriented as he let down toward what he thought was the Newark Airport; he may have mistaken the East River for the Hudson. But he was over midtown Manhattan and the rest came quickly. For the B-25 Mitchell it was a major headline though of a kind one doesn't like to see.

Another spectacular crash had occurred the previous summer at Los Angeles involving a B-25. A standard B-25H (43-4406) had been selected for project 98X, an attempt to up the speed of the Mitchell. The Wright Cyclones were replaced with larger Pratt & Whitney R-2800's, each belting out 2,000 horsepower. Cowlings were taken from a Douglas A-26 to accommodate the big engines and large spinners were fitted to streamline the three-blade propeller hubs. The tests began, pushing up the speed a little at a time. Finally, on a high-speed pass over the airport—probably approaching 400 miles per hour—the main spar failed and the fuselage shed its wings and continued into the ocean at the runway's end. Everyone aboard was killed—a high price sometimes paid in order to improve a design. That project was abandoned. Other times an experiment may not pay off and be costly only in time and effort. Such was a modification of a B-25C to four-bladed props and the short Clayton exhaust stacks. No improvement in performance could be found.

In March 1945, at the request of the U.S. Army Air Force Material Command, a B-26G was taken off the ready line for a special mission. It was to find what a bicycle-type landing gear would mean on a modern high-performance, relatively heavy aircraft. Back in the stick-and-wire pioneering days this system had been used on occasion, not so much for drag reduction, but more because it was just another way of attaching an undercarriage to a flying contraption. It really had no precedent to live up to. Hubert Latham's *Antoinette IV,* lost in an English Channel crossing attempt the week before Bleriot's success, was one of the more famous users of this type gear. What brought the concept out of mothballs in 1945 was the advent of the turbojet and the problems foreseen by the Air Force in trying to build a multi-jet bomber. Main landing gears had always before retracted nicely into the nacelles behind the engines. However, the jet engine was long and cylindrical with an inferno of hot gasses spewing out its tail. That was certainly no place to retract a landing gear. True, special pods could be constructed on the wings to house it, but that meant increased structural weight and more drag, although on many bomber designs the Russians have taken this approach. Ruling that method out, only the fuselage was left and that's what the Army Air Force wanted to test. There were possible problems with the system. What about differential braking—or, for that matter, differential throttle—during ground handling and steering? What about landing flare and rotation at takeoff? These were unanswered questions, along with cross-wind landing problems that might arise. The modified aircraft's designation was changed to XB-26H and it was colorfully dubbed the *Middle River Stump Jumper.* The Martin plant was at Middle River and the Stump Jumper part came from the four degree ten minute attitude the ship had on the ground. It looked poised and ready to leap into the air on its bicycle gear—and for good reasons. Without being able to rotate on takeoff, Martin engineers built the takeoff angle right into the landing gear.

The "Middle River Stump Jumper" shows off its bicycle gear in this head-on shot. The gear, for landing and takeoff tests only, did not retract. (USAF)

Looking like a giant insect about to leap into the air, its easy to see how the "Jumper" acquired its name. Only one of these aircraft was built. (USAF)

This aircraft was redesignated from B-26G-25-MA to XB-26H-MA when it was converted to a landing gear test aircraft. The 3.5-degree upward canted engines and wings show up well in this photo as well as the fuselage strengthening strakes. (USAF)

The "training wheels" had a sort of knee action arrangement for long travel which allowed to some extent for a wing low landing. (USAF)

A quick walk around showed up the differences between the *Stump Jumper* and a standard B-26. Obviously, the tricycle gear was gone, now replaced by the fore and aft main wheels under the fuselage. The struts were mounted so that the aircraft's weight was shared equally by each. They were standard B-26 shock struts located on the fuselage center; they had formerly been on the nacelle center line. The forward gear was mounted in the aft section of the original nose gear well and the rear member was positioned in what had been the rear bomb bay in earlier models. The wheels were 17.00 x 20 and the forward gear was able to turn 30 degrees in either direction for steering. Like the B-25 nose gear, it was spring loaded to return to a straight-ahead position when it was fully extended with all weight removed from it. Of course, being able to turn, a shimmy damper had to be built into the strut. Since steering could not now be accomplished with foot pedal brakes, the brake control was moved to a hand lever near the pilot's seat. However, steering the nose wheel was unique at best. A system of foot pedals translated the pilot's wishes to the front wheel. Several investigations of pedal travel versus pedal pressure were evaluated, with the travel system winning out as most comfortable for the pilot. The main gear, that is the bicycle wheels, did not retract, as the purpose of the test was ground handling and landing and takeoff studies, none of which were in the high-speed flight regime where streamlining was important.

To prevent the aircraft from tipping over on the ground an outrigger gear had to be fitted—much like training wheels on a child's bicycle. This was fairly easy as the engine nacelles were there and the main gear struts were hanging down from them. Twenty-six-inch wheels were attached to the strut axles and there you were—instant training wheels. They had 3½ inches of ground clearance. This gave the pilot an odd feeling as he turned in one direction but tilted or banked in the opposite when taxiing. After experimentation the struts were lengthened and the condition was somewhat improved. Since the weight of the aircraft was now being supported by an area of fuselage structure it was not designed for, the fuselage had to be suitably strengthened. A series of eight external stringers evenly spaced on either side for a goodly part of the fuselage length did the job.

On May 4, 1945, the first flight was made and the test series continued through June 4. Martin's test pilot, E. R. "Dutch" Gelvin, was selected to conduct the anticipated rough test program. Things like landing rear wheel first (which seemed okay), and landing forward wheel first, which produced an initial bounce that quickly accelerated into a porpoising nightmare, were on the schedule.

Another jaw-rattling maneuver was a full stall a few feet above the runway, the B-26H dropping straight in as a test of the pilot's behind and the shocks. After Gelvin emerged relatively unscathed from the program the XB-26H was flown over to Wright Field for some service pilot evaluation. Twelve Air Force pilots checked it out while bearing in mind the end usage in the next-generation jet bombers. Generally their response was favorable.

After the official USAAF trials, the *Middle River Stump Jumper* returned to Middle River and Martin where it was once more put in the hands of Gelvin. They fitted the gear up with strain gauges and told him to break it up if he could. He tried the nose-wheel-first landings, cross-wind landings and taxiing at high speed while making "S" turns. Gelvin lived and so did the gear! The data then belonged to the aero engineering departments across the land due to the contractual nature of the program with the USAAF.

Martin immediately grabbed the idea and applied it to their XB-48, a six-jet medium bomber with straight wings and tail. Boeing liked it, too, with the bicycle gear. With their XB-47's long, slim cylindrical fuselage, and thin-airfoiled swept flexible wings with six jets in pods slung underneath, the bicycle gear was a natural and they quickly adopted it. They also beat out Martin for the contract for the major Strategic Air Command (SAC) bomber of the mid-fifties. They went into production and proceeded to build hundreds of B-47's for America's first line of defense. So much for stump jumping, except it should be pointed out that this same type landing gear flies today with SAC's B-52 and the stealthy, slippery Lockheed U-2 spy plane. Another headline? Yes, it was shown in newsreels laying down rubber as it made it's "S" turns at high speed for the fortress of democracy.

Both the B-25 and B-26 had movie careers and each played their roles co-starring with Spencer Tracy and Van Johnson, though in different movies, both MGM pictures shot during the war. *Thirty Seconds Over Tokyo*, the 1944 recreation of Doolittle's epic raid of early 1942, was probably as accurate as any aerial war movie ever made. The event was fresh in everyone's mind, and the equipment was still available and in first-line service right down to the proper models and markings. The B-26 was not cast quite so accurately. In the movie *A Guy Named Joe*, about half of the Marauders of the 21st Bomb Group at Tampa's MacDill Field were painted up to look like the Mitsubishi G4M's. Those were the Imperial Japanese Navy's land-based medium bombers which we code-named "Betty." Large orange-red rising sun insignia's were painted on the fuselage sides and wings along with suitable Japanese alphabet characters. A white stripe was painted around the fuselage just in front of the fin and three white stripes were painted across the rudder. The effect could almost make you believe they were Japanese bombers and, since the film was shot in June and July 1943, MGM wasn't about to get a contract with the Japanese for use of the real thing. They were fitted up in some cases with smoke canisters slung under the fuselage so they could go down in flames (at least smoke) whenever Van Johnson would shoot 'em up with his trusty P-38. It was all for the war effort and soon thereafter the crew members that took part in the movie found themselves with the 322nd Bomb Group in Europe flying in the real thing.

It wasn't until the late 1960's that the by then old B-25's were hauled out of mothballs for yet another starring movie role. This time it was Joseph Heller's super-novel *Catch 22*, with its less than enchanted view of life in a B-25 outfit in Italy during World War II. With its wonderful characters—Yossarian, Colonel Korn, General Dreedle, et al—it was a fine movie. Who can forget the opening scene of total darkness, a dog barking far away, the blazing first rays of sun, and the cough of R-2600 Wrights coming to life—the prelude to eighteen B-25's in a mass takeoff. To get to that point required three years of Frank Tallman's best efforts in gathering eighteen assorted ragtag B-25's from around the country, nursing them carefully back to California and rebuilding them into bombers once more. In addition, crews had to be trained to bring them up to the flight proficiency of a real combat unit. The film was shot in Mexico where the arid countryside was similar to southern Italy. It was one of the most famous squadrons of B-25's ever to fly in the movies. The most recent use of these aircraft in the movies was in 1978. Five B25's were flown from the U.S. to Luton in the U.K. repainted drab with ficticious serials and used in the film 'Hanover Street.'

Another area of interest with respect to the B-25's was its continued postwar use by the Russians. Under lend-lease the United States had, from time to time, furnished the USSR with Mitchells of all models—B's, C's, D's, G's and J's. Just as the postwar USAAF continued to use them in a variety of roles, so did the Russians, well into the 1950's. Theoretically, now it was a Soviet bomber and therefore had to be given a NATO code name. So joining the "Bisons," "Bears," "Backfires," etc. were the hapless Communist Mitchells—code name "Bank." There were also a few Red Chinese Air Force units equipped with Mitchells captured from the Nationalist Chinese.

After World War II, in many Latin American countries it became popular to have an air force. With so many surplus B-25J models hanging around it was not long before they were either given to or bought by these countries. Bolivia, Brazil, Chile, Cuba, Mexico, Peru, Uruguay, Venezuela—all procured them in various quantities. Indonesia was later added to the roster of foreign countries. As late as 1963, Brazil, Chile, Peru and Indonesia were still fielding the ships in bomb groups they considered to be their first line of defense.

The B-26 disappeared rather quickly from the scene after the hostilities ended. France maintained a few squadrons for a short while but the jet age soon overtook the Marauder in the bomber role. In fact the French did use one B-26 for a jet engine test bed complete with air scoops on the fuselage sides and tail pipe out the rear. Since they weren't as easy to fly not many were snapped up for the roles that allowed the B-25 to continue serving for years to come. Most were destroyed in Europe by dynamite blasts rather than even bother to bring them back to the United States. It's just possible that even today, or tomorrow, the Mitchell may make one more headline. There are still several of them out there flying around, fighting forest fires as water and chemical bombers, as executive transports, and as rich men's playthings.

Back when the Russians were our World War II allies, we provided them with a large quantity of armaments including B-25's. They flew crews into Alaskan airfields by transport to take delivery of the aircraft. This B-25C is being picked up after being painted with USSR markings in the late summer of 1942. (USAF)

8

MANICURED MEDIUMS—THE BUSINESSMAN'S BOMBER

AT THE END OF WORLD WAR II, American industry began to pick up the pieces and return to a peacetime economy. No more "A" gas rationing stickers on windshields, no more shoe stamps, no more food tokens. All the money that Rosie the Riveter and her fellow civilians had put aside during the hoopla of endless War Bond drives was now out there waiting to be spent on goods that had been totally unavailable for four years.

Rising to the occasion, as they always do, captains of industry, with the old profit motive firmly fixed in mind, took advantage of every short cut possible to get more productive work packed into their days. America had come to accept the airplane matter-of-factly, unlike the "gee whiz" attitude that had pervaded before the war. Airlines were now scurrying to get their well worn aircraft back from the armed forces, but their schedules were still relatively meager. The businessmen, the top corporate executives, needed some type of high-speed, long-range transportation for their headlong scramble to restore the postwar world of commerce and industry.

The answer was military aircraft conversions, and for pure speed, pizzazz and comfort what better than a sharped-up B-25 or, in a few cases, a B-26. In both aircraft the fuselages were roomy enough to gut the bomb racks and other military hardware out and replace them with interiors even an oriental potentate would turn an envious green over. Fancy upholstery lined the cabin walls, large picture windows were cut in the fuselage flanks, rich desks, chairs and a washroom, bar and galley completed the lavish refurbishments. The flight decks were just as elaborate in their own way, sporting the latest in instrumentation to make sure Mr. Big got to his destination safely and on time. The engines were totally overhauled or replaced with new and sometimes larger ones. With the executives willing to loosen the corporate pursestrings for their own comfort and safety, there was no limit to what could be done to make these planes the absolute cream of the business fleet.

The B-25's, and to a lesser degree B-26's, successfully served this purpose until the advent of the business jet, such as the Lockheed Jetstars in 1960 and the later Learjets. Only at this point would the manicured mediums be beaten at their own game of speed, safety and range. A whole new industry had developed to modify, recondition and maintain these refugees from mothballs and the scrapheaps they were headed for after World War II. Many light and medium bombers were declared surplus and snapped up by the rebuilders. Before long the number of corporate executive transports far surpassed the airline transports in service.

Favorites of the corporations varied widely from *The Chicago Tribune*'s Colonel R. R. McCormick's Boeing B-17, to Coca Cola's R4D-1 (Navy DC-3); from the Tennessee Gas Transmission Company's B-26's refitted with DC-6 engines and propellers and with a range of over 1,500 miles, to the long-range spruced-up Timkem Roller Bearing Company's B-25D that seated eight and sported Patuskin tip tanks which improved the wings' efficiency. That plane even had an electric wheel chair elevator in the aft compartment. Other aircraft ideally suited for the hot rod executive peacetime conversion were the Lockheed C-60 Lodestar, the Consolidated B-24, the Douglas A-20 and A-26 and the Douglas B-23. Marathon Oil, Continental Can Company, Hughes Tools Company, Standard Oil, Thatcher Glass, National Homes and hundreds of other corporations flocked to buy the ultimate in corporate status symbols—the corporate executive transport.

N171E and N5546N were sister ships belonging to Tennessee Gas Transmission Company and modified almost beyond recognition from standard B-26's. The craft featured air stairs, picture windows and powerful Pratt & Whitney Twin Wasp R-2800-CA-15 2,100-horsepower engines. The engine prop combinations were taken from Douglas DC-6 transports. Range was over 1,500 miles. (Collect Air Photos)

More beautifully modified B-26G's for the business fleet during the early fifties. (Collect Air Photos)

North American Aviation was way ahead of them. NAA test pilot Vance Breese had poked the first B-25 (40-2165) upstairs on August 19, 1940, for its maiden flight. Being the only test aircraft of the new design, it went through the vertical tail modifications and trials and the eventual change to no dihedral on the outer wing panels to prevent the longitudinal instability (dutch roll) that was the prototype's only defect. After many tests and changes during the test program the prototype became a little shop worn, out of date, and of little use to the military. It then sat around the NAA ramp for awhile as the war production effort was cranked up to full steam. By this time the NAA Kansas City and Dallas plants were under way and Dayton's Wright Field, Washington's Pentagon and numerous suppliers were on the list of places to be visited by the NAA brass. With the airlines at a real low on planes available (the USAAF and USN had taken over a large share of the fleet for their own purposes), the mode of transportation was not much better than rail or bus travel. No-nonsense Dutch Kindelberger, NAA's president, grabbed the B-25 ramp sitter and had it wheeled into the shop. There it was stripped of her military heritage, fighting gear and all, and refitted as a flying Taj Mahal for the NAA execs. Carpeting installation, upholstery, desks, intercom, washroom and a bunk over the former bomb bay added luxury to the seven passenger seats installed, two in front of the bomb bay and five aft. The bomb bay itself made a good place for the baggage. The bombardier's greenhouse was scrapped and an aluminum skin covering was designed, formed and placed over it—a modification not to be seen again until the final J model, three years later. The aft cockpit windows were also covered and four windows were installed on the fuselage sides for the passengers.

Due to a small portable bar—never officially spotlighted by North American—the aircraft was nicknamed "The Whiskey Express." Its pilot, Edgar "Ed" Stewart, probably spent more hours at the controls of a B-25 than anyone else ever as the executive transport plied the routes of wartime aerial commerce.

Throughout the war all went well with B-25 Number One until January 9, 1945. What goes up must come down, and unfortunately sometimes pranged. Pilot Stewart, copilot Theron Morgan and crew chief-engineer Jack Maholm were checking out a routine engine change. Winter at Inglewood's Mines Field was usually a little soupy and that day was no different. An engine went out and Stewart made his approach about 40 miles per hour faster than the book speed, just to be safe. However, what he didn't know was that an orifice in the hydraulic landing gear system had been drilled oversize and when he tried to extend the gear at the overspeed it would not lock in place. By now he had slowed down, used up too much runway, and decided to go around. No way! He was out of airspeed, altitude and ideas all at the same time and the B-25 plopped in. No one was hurt but Numero Uno was scrapped and carted off to the smelter.

The Southern Minnesota Wing of the Confederate Air Force has this pristine B-25J as a showpiece at air shows. (Collect Air Photos)

These photos show what time, money and desire could do for the B-25 once conscripted into the businessman's bomber squadron. (Collect Air Photos

Up to 1949, six B-25's were converted by NAA to what were redesignated RB-25's; in actuality executive transports, these predated the conversions of the late forties and early fifties for the corporate execs. The first, of course, was the *Whiskey Express* just mentioned. Next came General Hap Arnold's aircraft. He showed up at Inglewood one day and while there on another mission lustfully cast his eyes on Kindelberger's B-25 *Whiskey Express*. Well, being Air Force Chief of Staff, he wanted one too. It was mid-1943 and production shortages were no problem, so he had a B-25C diverted to the Field Service Department for a general fixing up to executive transport standards, like 40-2165. The government paid for it and, eventually, after the war, Howard Hughes latched onto the aircraft and used it on and off for another twenty years!

Timkem Roller Bearing Company's B-25D-NC conversion was a real beauty and first-class executive transport. Features included Patushkin tip tanks that added range and greatly improved the wing efficiency. The aircraft seated eight in luxurious comfort. To keep cabin noise down, it was fitted with a full collector ring exhaust system. An unusual oddity was the electric wheel chair elevator fitted in the rear fuselage. (Collect Air Photos)

CB-25J was General Dwight Eisenhower's executive transport. Starting life as a B-25J-1-NC, it was modified by North American for the Supreme Allied Commander. General's blue replaced the olive drab fabric used to line the aircraft. At the time this photo was taken, the craft had been relegated to a mere two star general. (Collect Air Photos)

The next customer was none other than General Dwight D. "Ike" Eisenhower. March 31, 1943, a B-25J-1-NC was grabbed off the line at Kansas City and flown off to Inglewood on a "secret" mission. The conversion was completed May 11th. The interior lacked the quilted olive drab military upholstery, but instead had a nice "general's type" blue fabric. Clamshell doors were fitted to the nose for navigation and radio equipment easy access. The entry hatch at the rear (the one usually used by the rear turret gunner and waist gunners) was moved back to obtain more floor space for chairs and a drop-leaf desk for the General. Like a DC-3, luggage racks were put above the seats, and the bomb bay was taken up with long-range fuel tanks in the upper portion and additional luggage space below. The craft started out being designated RB-25J, then went to a CB-25J designation. The AAF serial was 43-4030. The bombardier's position was covered over and extra windows were added to the fuselage sides. Eventually, a major general inherited it from Ike.

Albert Trostel & Sons, Company, tanners from Milwaukee, sported this "J" fixed up for business. Licensed N75754, it was a sister ship to N32T. (Collect Air Photos)

Another beauty that worked at the tannery, N32T shows the bay window waist gunner's position being used for just that—bay windows! These planes were B-25J-35-NC's. (Collect Air Photos)

Come 1944 and General Arnold was ready for a new one—his B-25C was dog-eared by now and he opted for a J—just like buying a new car every couple of years. Out of the hopper came a spanking new RB-25J-15-NC, AAF 44-28945, and it was turned out almost identical to Eisenhower's exec hot rod of a few months before. Hap used it till after war's end in 1946. Like Ike's steed, it was durable, and after the General got through with it, it was still used by colonels well into the 1960's.

When NAA's executive plane pranged out they of course needed a replacement—their brass had gotten used to being whisked (whiskied?) about the country in their fast, high-powered aircraft rather than relying on the still rickety scheduled airlines. They took a J model in January 1945 and began a conversion similar to the Eisenhower and Arnold J's. However, they decided to do a bit more on the craft's civilianization. After all, with the war now nearing conclusion there was the possibility of selling the B-25 design in the post-war corporate market as executive and light transports. A heater was added in the radio bay along with a radio directional finder. A solid-nosed J model was used, with zero hours on it. By March of 1945 the aircraft was completed and put in service by NAA. Between company trips they flew CAA certification tests with it in hopes of clearing it with the government for future civilian sales. On February 27, 1946, during a test over the Pacific, test pilot Joe Burton radioed in a mayday—a wing was on fire. Shortly thereafter, the aircraft exploded, taking the lives of the crew and falling into the ocean.

That was the end of the conversions for awhile, until 1950, when North American purchased a surplus B-25J and took another shot at an executive transport. This time they built a new nose for the ship, extending the overall length about 40 inches. At the same time the nose was about 14 inches wider, giving a 70-inch cockpit width. That allowed the pilots' seats to be moved forward and four passenger seats to be mounted between the bomb bay bulkhead and the pilots. It was a good arrangement, even for military training use, for three students, instead of one, could be dealt with simultaneously. The bomb bay was permanently closed over as a baggage compartment, with a hatch and electric baggage lift near the front of the compartment. Since the nose was wider, a Convair 240 windshield was used. The upper short exhaust stacks were removed and a 180-degree collector ring installed in their place to reduce the noise level, which it did. Dubbed the *Bulbous Nose B-25* it obtained a civil registration number N-5126N and initial flight testing proved the aircraft completely airworthy—until March 25, 1950. The aircraft broke up in a flight through the extreme turbulence of a storm front near Chandler, Arizona, taking the lives of seven NAA people. With that the project was abandoned.

Another interesting conversion was made by Hollywood's Paul Mantz for movie filming. The nose of a B-25 was completely redesigned to mount the complex camera system used for Cinerama. That aircraft still exists, and is used today for the same job. Home base is Van Nuys, California.

Paul Mantz had a special camera nose built on this B-25H-5-NA to take aerial motion pictures for Hollywood. The craft has been all over the world on assignment. It even took Cinerama. Still in use, it usually can be found at work whenever the motion picture industry needs an aerial sequence. (Collect Air Photos)

B-25's, other than the NAA conversions, were prepared for the executive market by outside companies and snapped up by transportation hungry (and status symbol hungry) businesses. However, some 25's didn't quite get all the glamor treatment and wound up hauling meat out of the Brazilian back country to packers along the coast. Some became chemical and water bombers to control forest fires and even today are still doing that job. Also out in the boonies, on dark nights in Florida, one can hear the familiar noisy roar of a pair of Wright Cyclones dragging a B-25 along at full bore and low altitude. Only the sound is there as all lights aboard the aircraft are extinguished. It's a B-25 from the "Mary Jane" Air Force, plying its illicit dope trade from Columbia to Florida, or perhaps a load of firearms for a banana republic revolution on the way back to South America.

The B-26's, with few available, still retained their shady lady reputation for being a real hot ship, so it took a pretty flamboyant and gutsy board of directors to elect to buy one to haul themselves about in. However, some did and wondrously streamlined creations were accomplished such as the Tennessee Gas Transmission Company's super sleek duo. One B-26C owned by Allied Aircraft Company of North Hollywood, California, finally let its aircraft do what many people had thought appropriate ever since the basic lines were laid down—be an air racer. It was christened the *Valley Turtle* and drew race number 24 in the 1949 Bendix Trophy Race, a 2,008-mile event from Rosemond, California, to Cleveland, with a racehorse start. License number of the polished natural-metal Marauder was N5546N. They loaded up the bomb bay

Yes, there was a B-26 air racer. It was flown by Lee Cameron in the 1949 Bendix Trophy Race, a cross country event from Rosemond, California, to Cleveland, Ohio. A broken fuel line precluded its completion of the race. (Collect Air Photos)

with tanks and expected to fly the race nonstop. Pilot Lee H. Cameron had the honors and took off for Cleveland in good style, but had to set down in North Platte, Nebraska, to fix a fuel line. He continued on to Cleveland but finished after the race's deadline. Anyway, he got to see the rest of the races and had traveled from the West Coast in high-speed style.

It must be said that all medium bombers converted from war surplus to civilian use were not flying palaces. Some were real pigs, dented and with the license numbers scrawled on the sides with a sloppy dime-store paint brush without benefit of masking tape. The interiors were also in the "what you see is what you get" condition. The owners of these planes were usually small businessmen who had flown the ships during the war and still wanted to hell around in one after the war—when they could afford the gas. Also it *was* a tax write-off for their businesses.

Others, while not luxury liners, were still neat and well kept and owned by such companies as Gail Burchard Mobile Homes, Albert Trostel & Sons, Company (tanners from Milwaukee), and of course the beauty owned by the Southern Minnesota Wing of the Confederate Air Force. Husky Oil Company and Javelin also sported two beautiful B-25's. Even Jimmy Doolittle tooled around in Gulf Oil's B-25 after World War II.

The Air Force was not about to have its brass of the fifties fly around in just plain bomber versions of the B-25J's, of which it still had 1,400 on roster in 1950. While not as extensively modified as the old *Bulbous Nose B-25* that crashed, they still had a nice but military job done on them by either Hayes Aircraft Company of Birmingham, Alabama, or the Tucson, Arizona, branch of Hughes Tool Company. These conversions were really almost new airplanes when the work was complete. They had all-new wiring, hydraulic systems, radios and glass. The engines and props were usually new, or at least completely rebuilt. The fabric-covered control surfaces were stripped and recovered and new tires and brakes were included in the effort to make the B-25 born again.

A lot of these conversions by Hayes and Hughes became multiplace trainers for many uses, such as learning to operate the sophisticated radar fire control systems used on the new jet fighters, advanced pilot training and navigation training. After use in the Air Force, many of them went to Air National Guard Units, almost all of which at one time or another used the ship. Most of these were designated TB-25K, -L, -M and -N, dependent on when they were converted and for what use.

One executive has a different idea of what his business plane should be. Ed Schnepf, publisher of Challenge Publications of Canoga Park, California, has his *Executive Sweet*, a B-25J which has been restored as a pristine World War II bomber, complete with gun turrets, guns and ammunition, the latter static, of course. At the annual Gathering of Warbirds, and many other air shows, particularly on the West Coast, it is usually the hit of the show. The subject of several *Air Classics* articles through the years, it is a well known aircraft even to those who have never had the opportunity of seeing it in person.

One last civil conversion of the B-25 into a flying loveboat might be mentioned. It was the wedding present of millionaire Barbara Hutton to Proferio Rubriosa and, along with being generally well appointed, had a king size bed with suitable enclosure and a leather-chaired john fit for the richest Saudi Arabian oil magnate. Ah, love.

Today at the Reading Air Show one or two executive B-25's will still show up, painted up in fancy dress and still dutifully hauling their corporate cargo of top officers. Properly maintained and cared for, a few will probably still be around in the year 2000.

B-25's in postwar livery, kicking around with the USAF and RCAF, doing whatever jobs were asked of them. (Collect Air Photos)

Almost all Air National Guard Units flew the 25 at some point. Here is a typical representation of four states. (Collect Air Photos)

Drawings

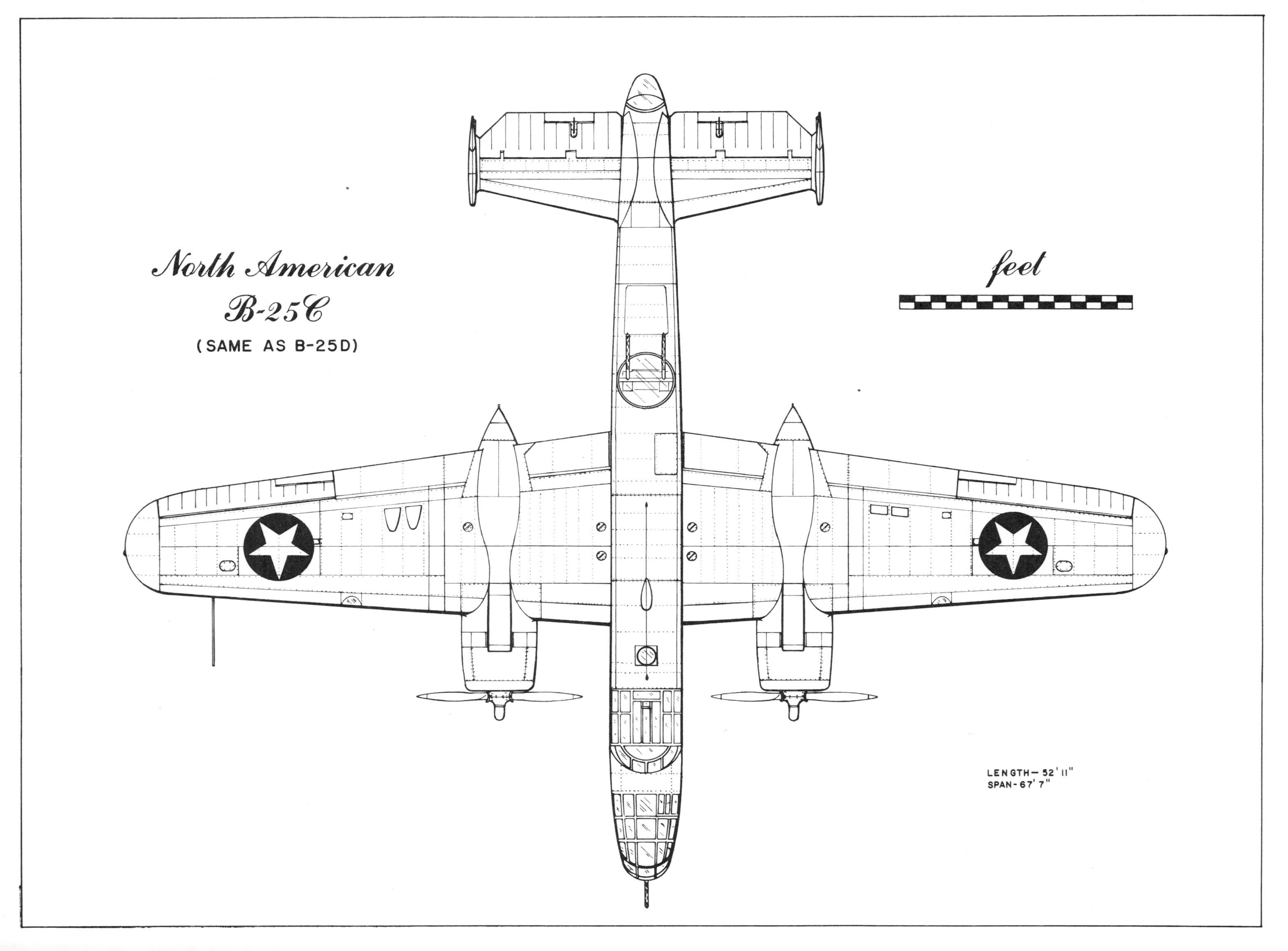
North American
B-25C
(SAME AS B-25D)
feet
LENGTH — 52' 11"
SPAN - 67' 7"

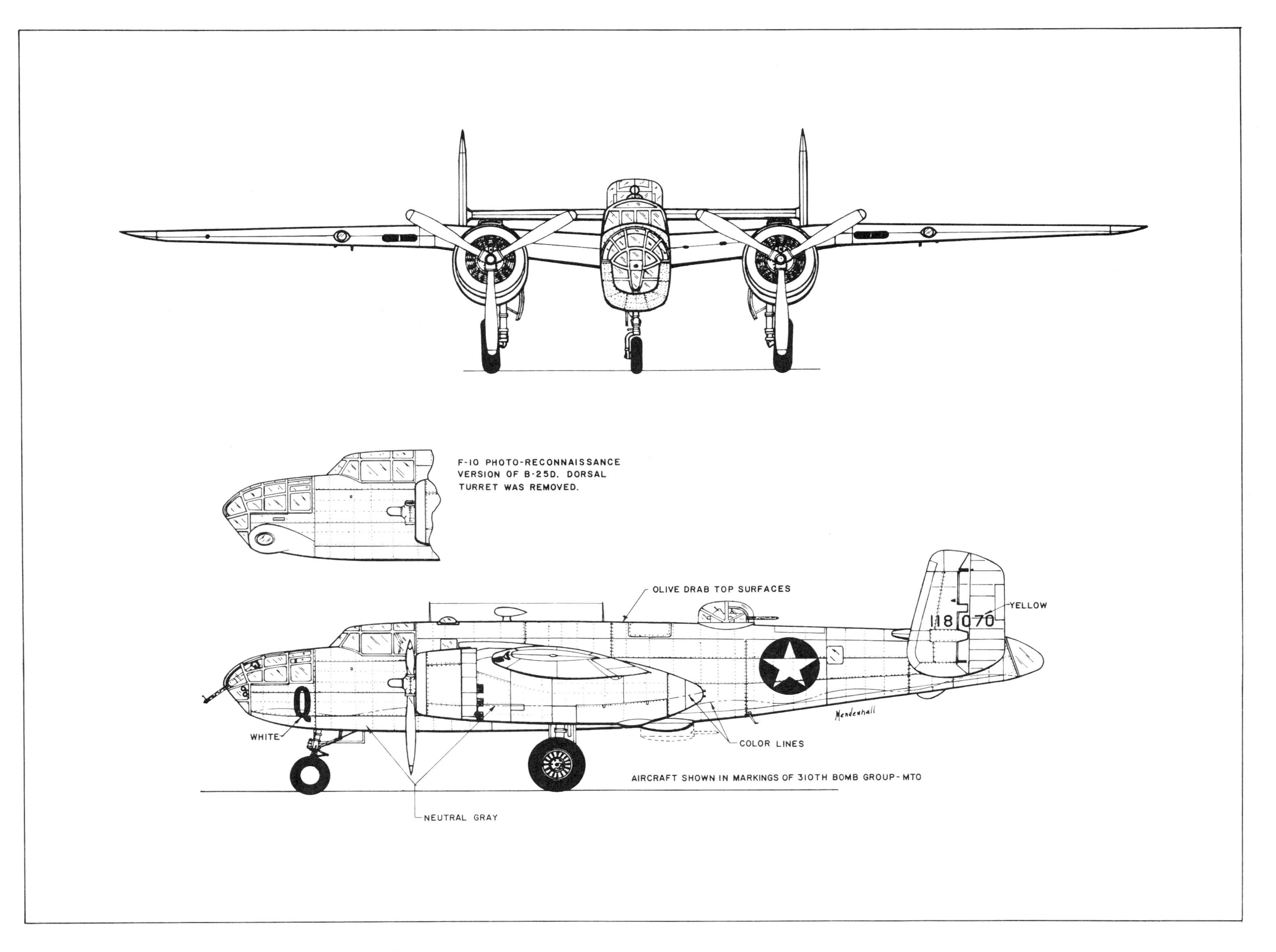
F-10 PHOTO-RECONNAISSANCE
VERSION OF B-25D. DORSAL
TURRET WAS REMOVED.
OLIVE DRAB TOP SURFACES
YELLOW
118070
Q
Mendenhall
WHITE
COLOR LINES
AIRCRAFT SHOWN IN MARKINGS OF 310TH BOMB GROUP - MTO
NEUTRAL GRAY

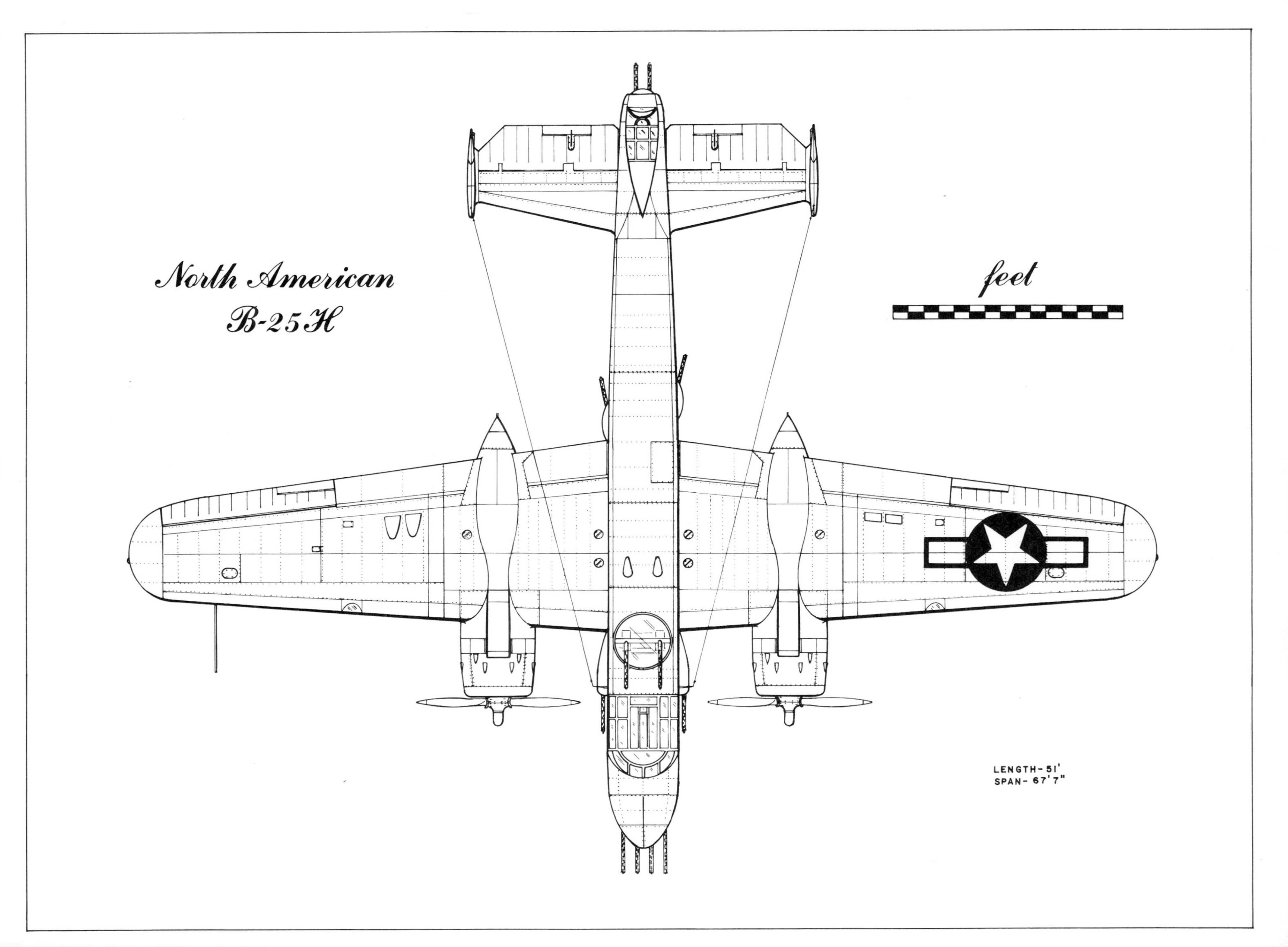

North American
B-25H
feet
LENGTH- 51'
SPAN- 67'7"

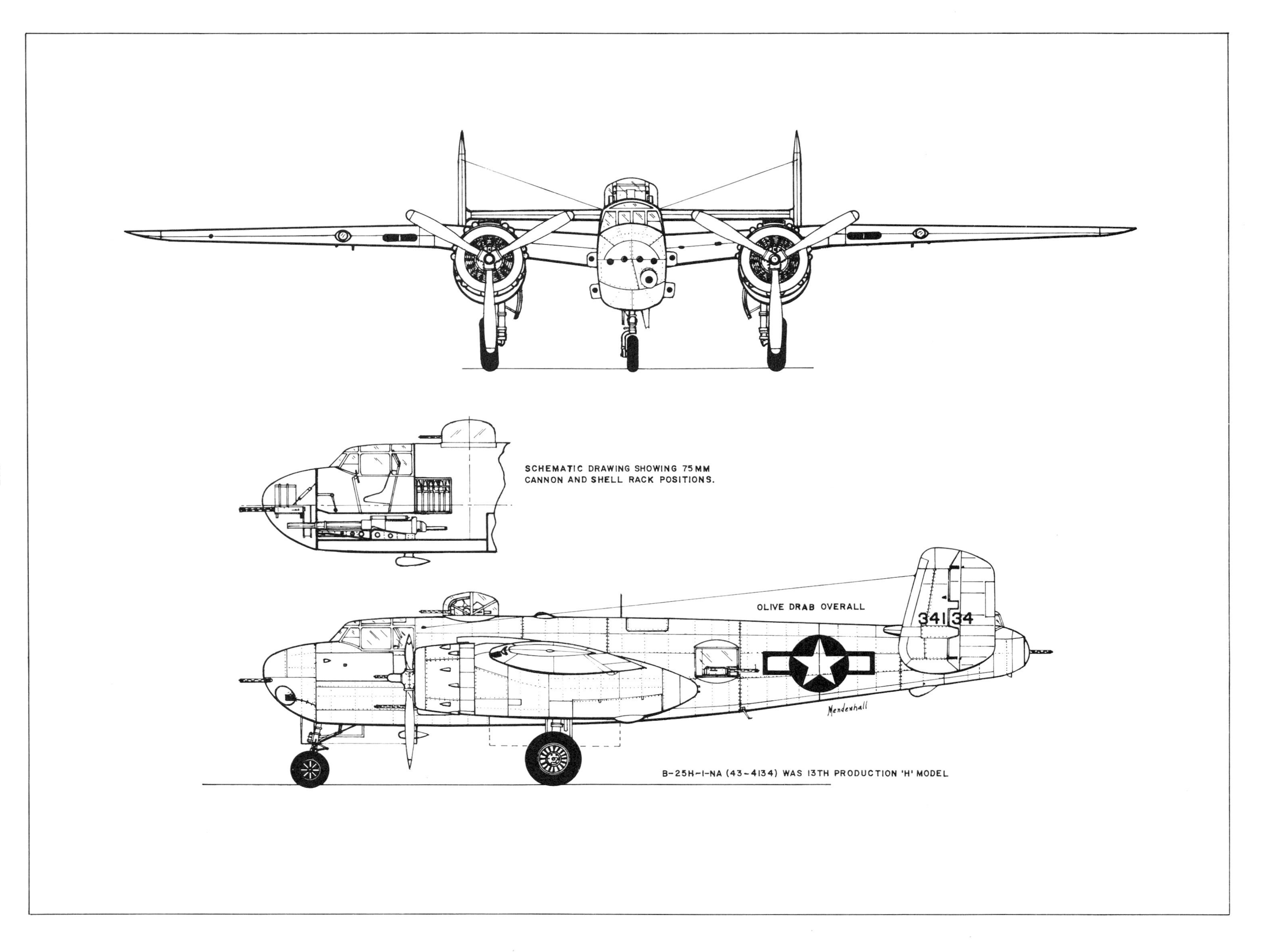
SCHEMATIC DRAWING SHOWING 75MM
CANNON AND SHELL RACK POSITIONS.
OLIVE DRAB OVERALL
341 34
Mendenhall
B-25H-1-NA (43-4134) WAS 13TH PRODUCTION 'H' MODEL

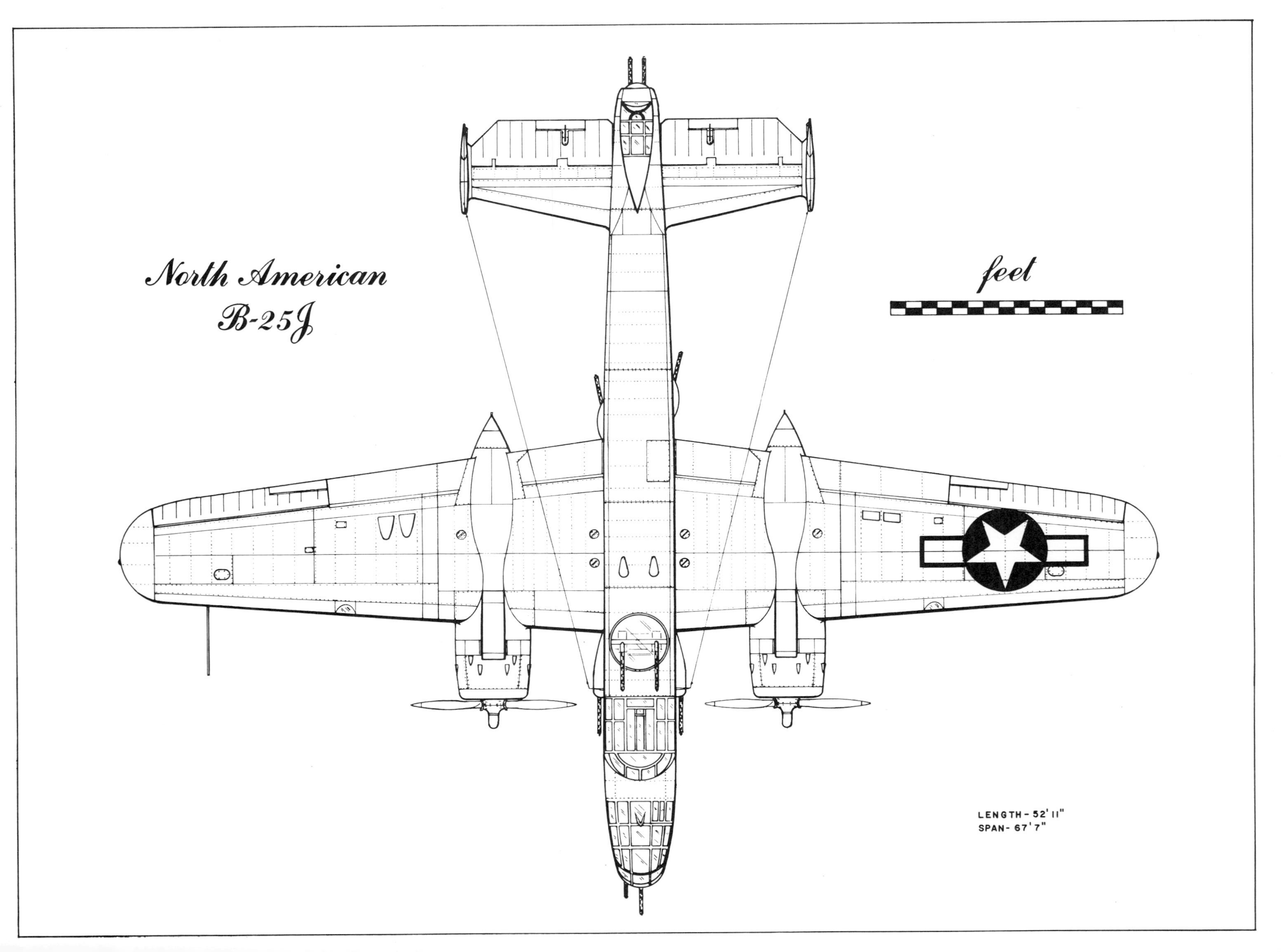
North American
B-25J
feet
LENGTH - 52' 11"
SPAN - 67' 7"

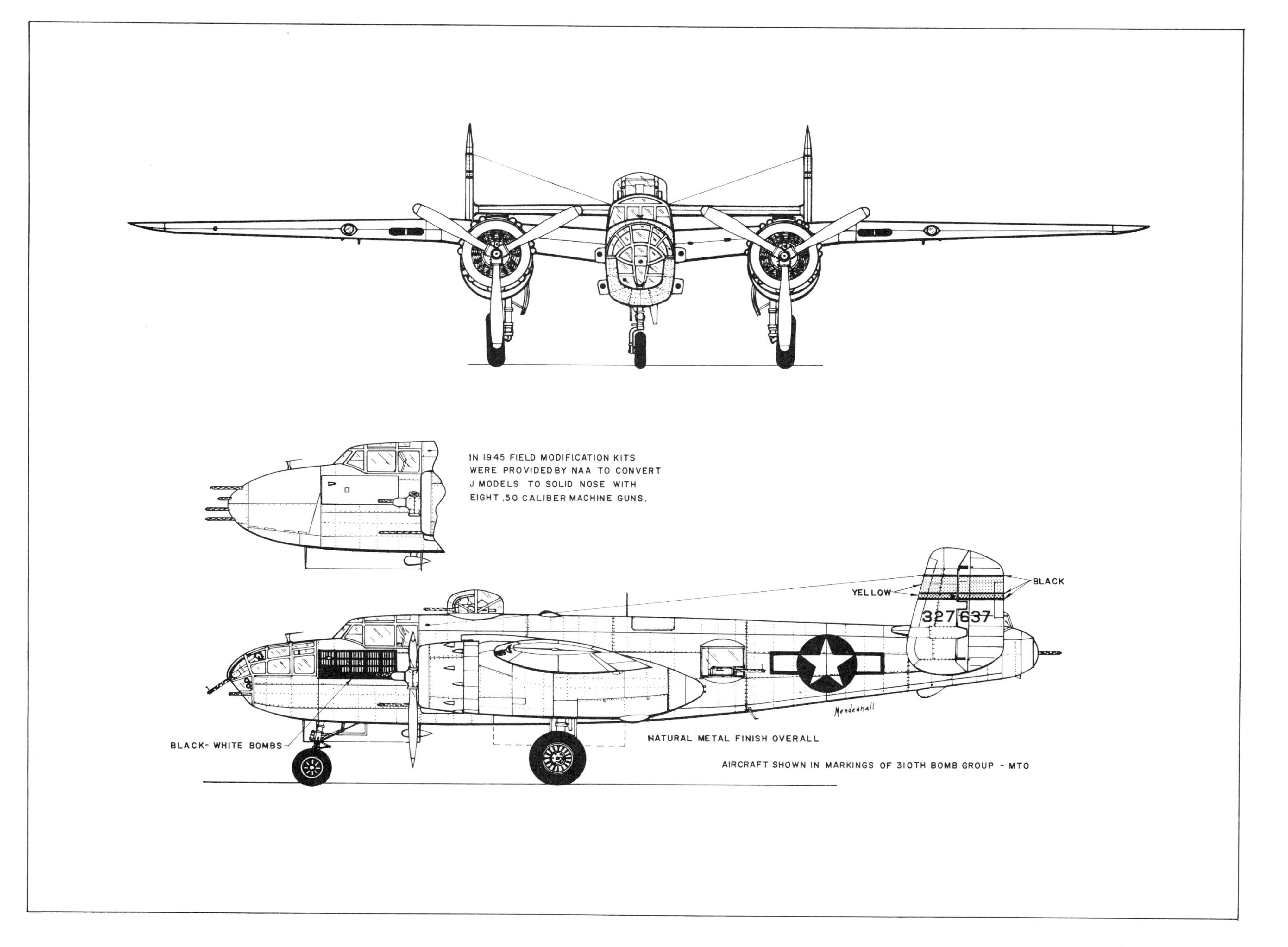
IN 1945 FIELD MODIFICATION KITS
WERE PROVIDED BY NAA TO CONVERT
J MODELS TO SOLID NOSE WITH
EIGHT .50 CALIBER MACHINE GUNS.
YELLOW
BLACK
327637
BLACK- WHITE BOMBS
Mendenhall
NATURAL METAL FINISH OVERALL
AIRCRAFT SHOWN IN MARKINGS OF 310TH BOMB GROUP - MTO

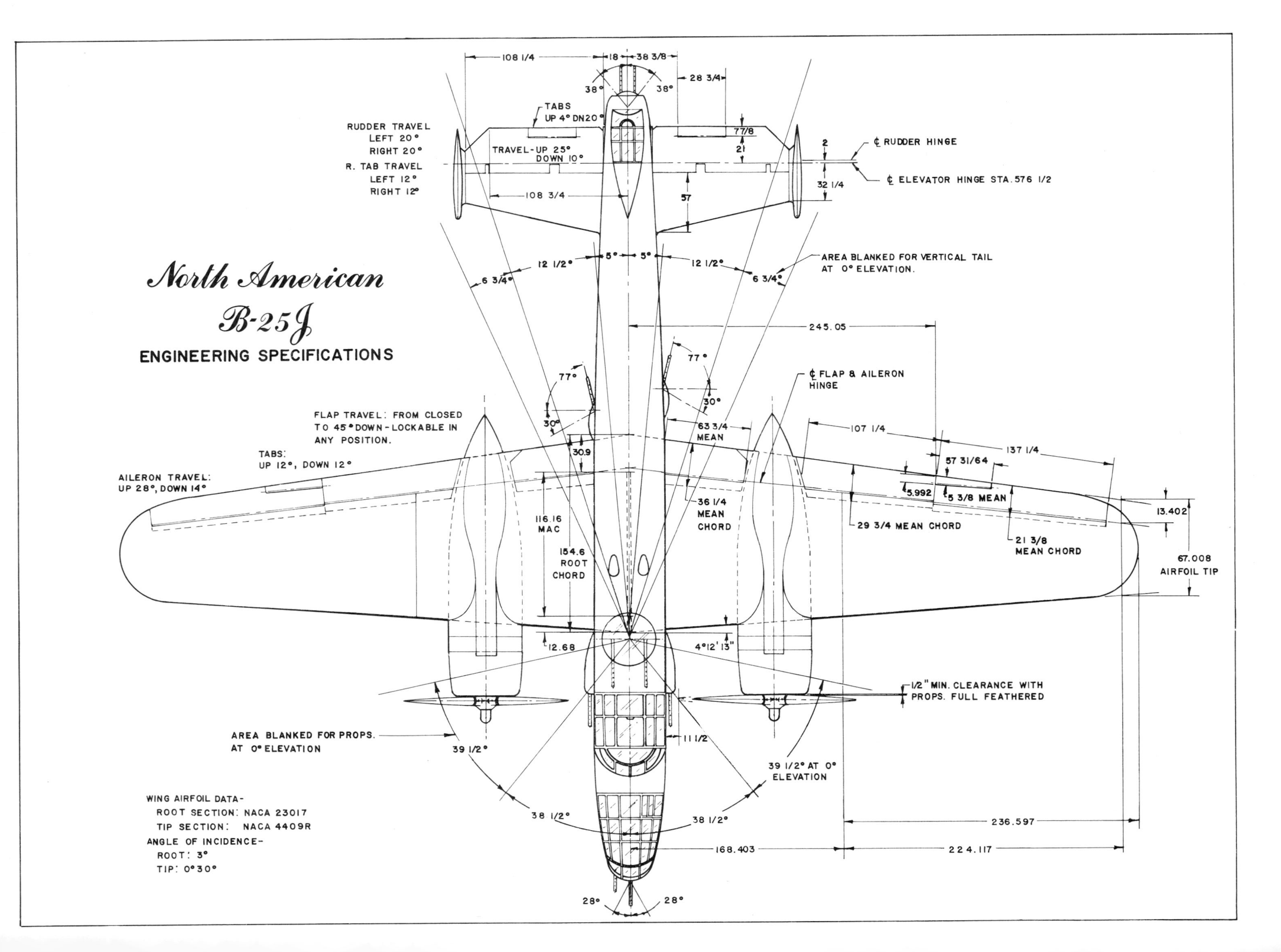
North American
B-25J
ENGINEERING SPECIFICATIONS
108 1/4
18
38 3/8
38°
38°
28 3/4
TABS
UP 4° DN20°
RUDDER TRAVEL
LEFT 20°
RIGHT 20°
R. TAB TRAVEL
LEFT 12°
RIGHT 12°
TRAVEL-UP 25°
DOWN 10°
77/8
21
2
℄ RUDDER HINGE
℄ ELEVATOR HINGE STA. 576 1/2
32 1/4
108 3/4
57
6 3/4°
12 1/2°
5°
5°
12 1/2°
6 3/4°
AREA BLANKED FOR VERTICAL TAIL
AT 0° ELEVATION.
245.05
77°
77°
℄ FLAP & AILERON
HINGE
30°
30°
FLAP TRAVEL: FROM CLOSED
TO 45° DOWN - LOCKABLE IN
ANY POSITION.
63 3/4
MEAN
107 1/4
137 1/4
TABS:
UP 12°, DOWN 12°
30.9
57 31/64
AILERON TRAVEL:
UP 28°, DOWN 14°
5.992
5 3/8 MEAN
36 1/4
MEAN
CHORD
13.402
116.16
MAC
29 3/4 MEAN CHORD
21 3/8
MEAN CHORD
154.6
ROOT
CHORD
67.008
AIRFOIL TIP
12.68
4°12' 13"
1/2" MIN. CLEARANCE WITH
PROPS. FULL FEATHERED
AREA BLANKED FOR PROPS.
AT 0° ELEVATION
39 1/2°
11 1/2
39 1/2° AT 0°
ELEVATION
WING AIRFOIL DATA-
ROOT SECTION: NACA 23017
TIP SECTION: NACA 4409R
ANGLE OF INCIDENCE-
ROOT: 3°
TIP: 0°30°
38 1/2°
38 1/2°
236.597
168.403
224.117
28°
28°

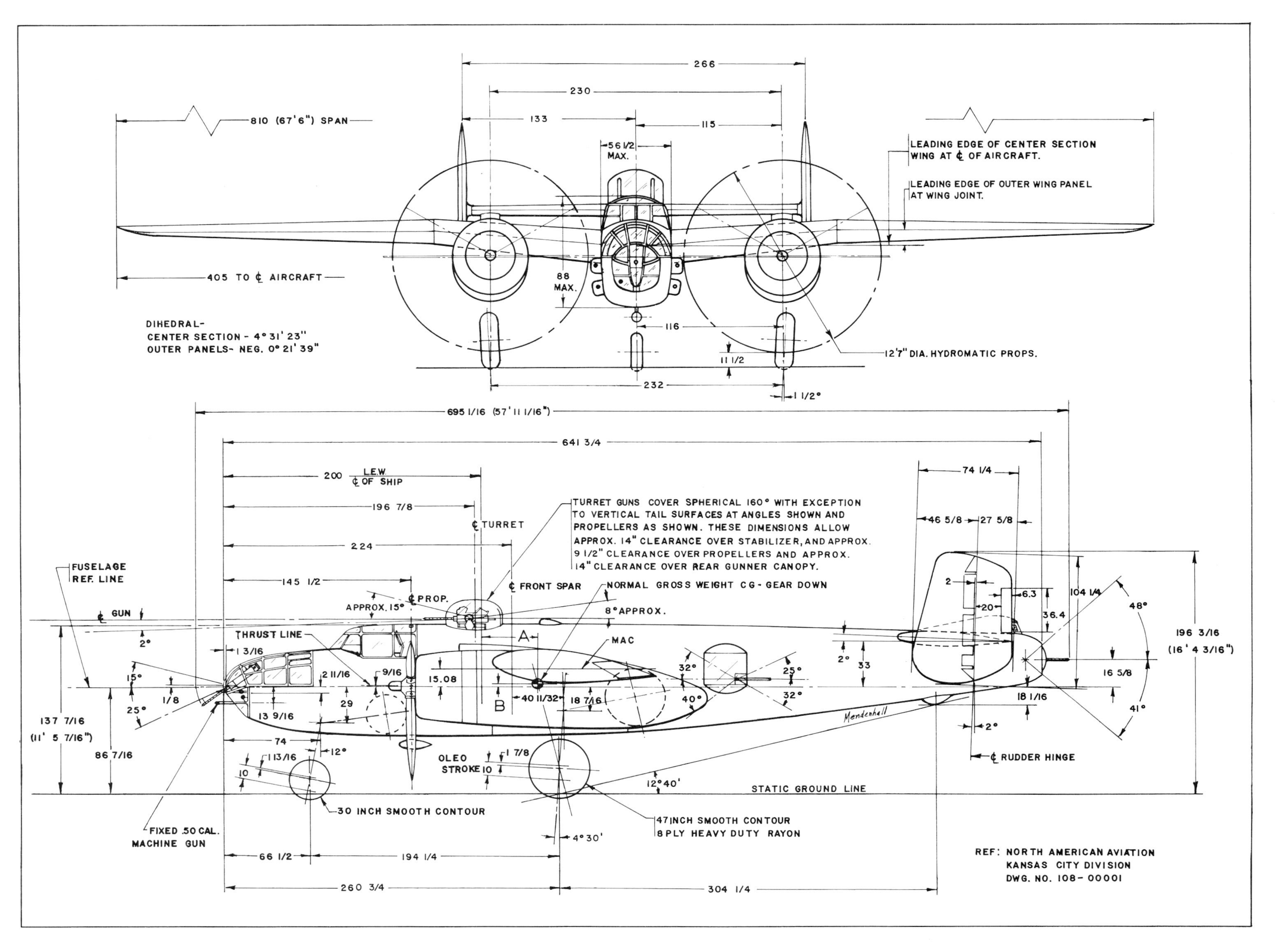
810 (67'6") SPAN
405 TO ℄ AIRCRAFT
266
230
133
115
56 1/2 MAX.
88 MAX.
116
11 1/2
232
1 1/2°
LEADING EDGE OF CENTER SECTION WING AT ℄ OF AIRCRAFT.
LEADING EDGE OF OUTER WING PANEL AT WING JOINT.
12'7" DIA. HYDROMATIC PROPS.
DIHEDRAL-
CENTER SECTION - 4° 31' 23"
OUTER PANELS- NEG. 0° 21' 39"
695 1/16 (57' 11 1/16")
641 3/4
200 L.E.W. ℄ OF SHIP
196 7/8
℄ TURRET
224
145 1/2
℄ FRONT SPAR
℄ PROP.
APPROX. 15°
TURRET GUNS COVER SPHERICAL 160° WITH EXCEPTION TO VERTICAL TAIL SURFACES AT ANGLES SHOWN AND PROPELLERS AS SHOWN. THESE DIMENSIONS ALLOW APPROX. 14" CLEARANCE OVER STABILIZER, AND APPROX. 9 1/2" CLEARANCE OVER PROPELLERS AND APPROX. 14" CLEARANCE OVER REAR GUNNER CANOPY.
NORMAL GROSS WEIGHT CG - GEAR DOWN
8° APPROX.
MAC
FUSELAGE REF. LINE
℄ GUN
THRUST LINE
FIXED .50 CAL. MACHINE GUN
OLEO STROKE 10
30 INCH SMOOTH CONTOUR
47 INCH SMOOTH CONTOUR 8 PLY HEAVY DUTY RAYON
STATIC GROUND LINE
℄ RUDDER HINGE
74 1/4
46 5/8
27 5/8
104 1/4
36.4
6.3
196 3/16 (16' 4 3/16")
16 5/8
18 1/16
48°
41°
137 7/16 (11' 5 7/16")
86 7/16
66 1/2
194 1/4
260 3/4
304 1/4
12° 40'
4° 30'
Mendenhall
REF: NORTH AMERICAN AVIATION
KANSAS CITY DIVISION
DWG. NO. 108- 00001

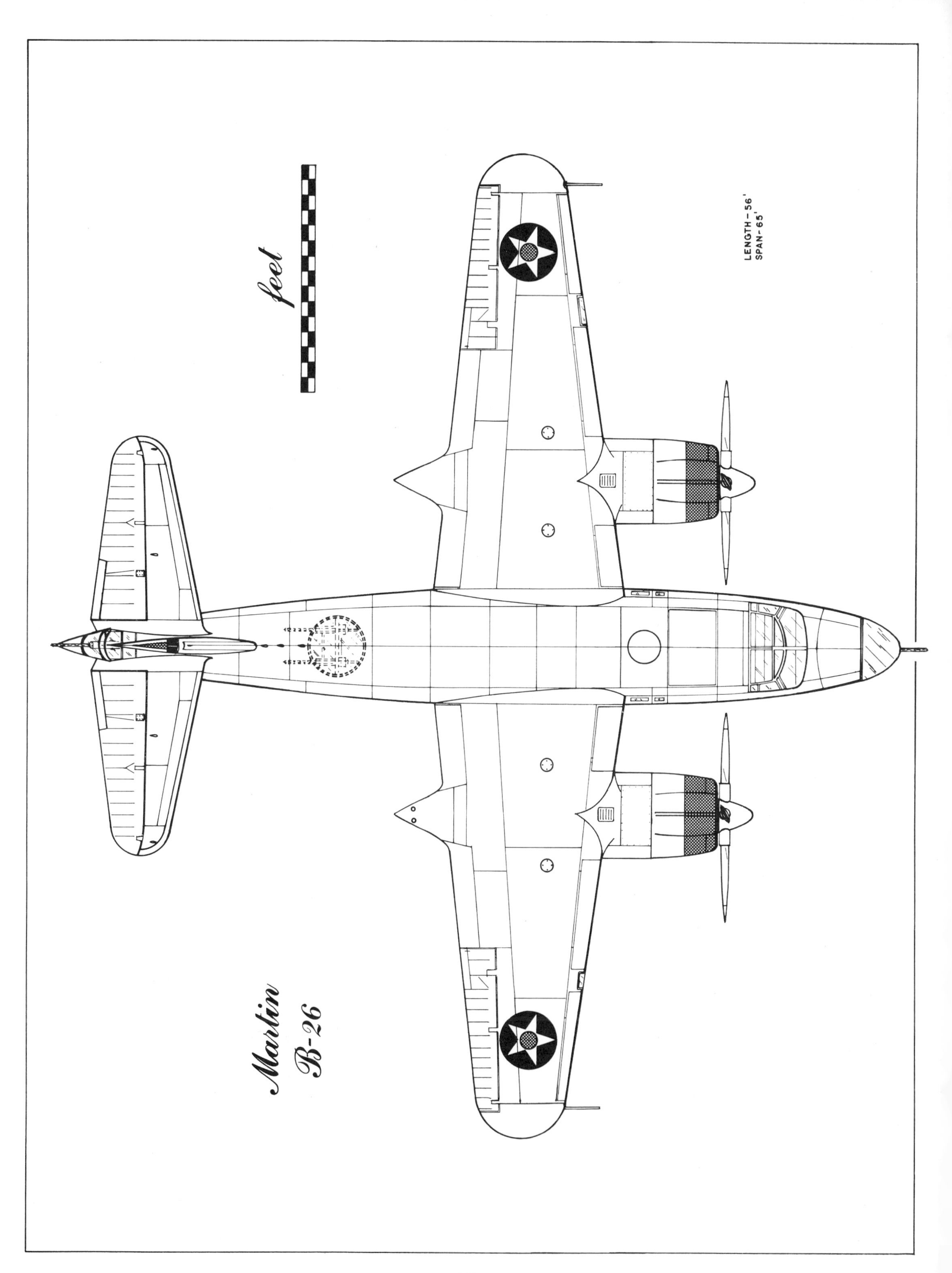
feet
Martin
B-26
LENGTH - 56'
SPAN - 65'

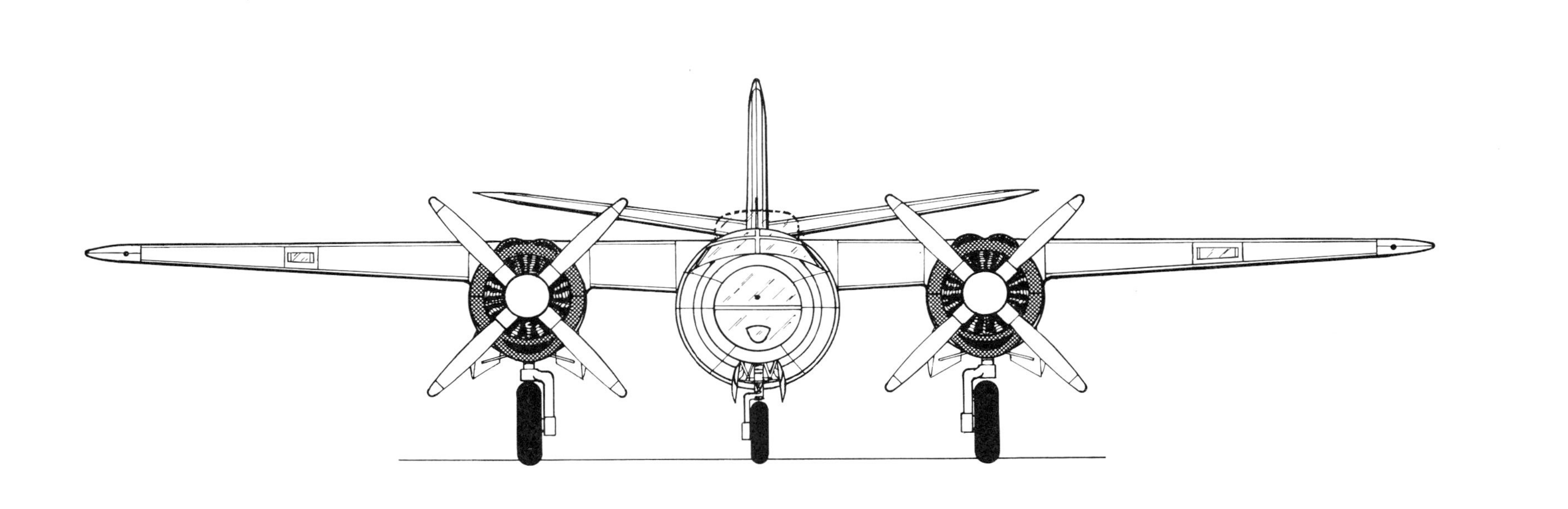

THIS AIRCRAFT, THE THIRD B-26 BUILT, HAD NATURAL METAL FINISH.

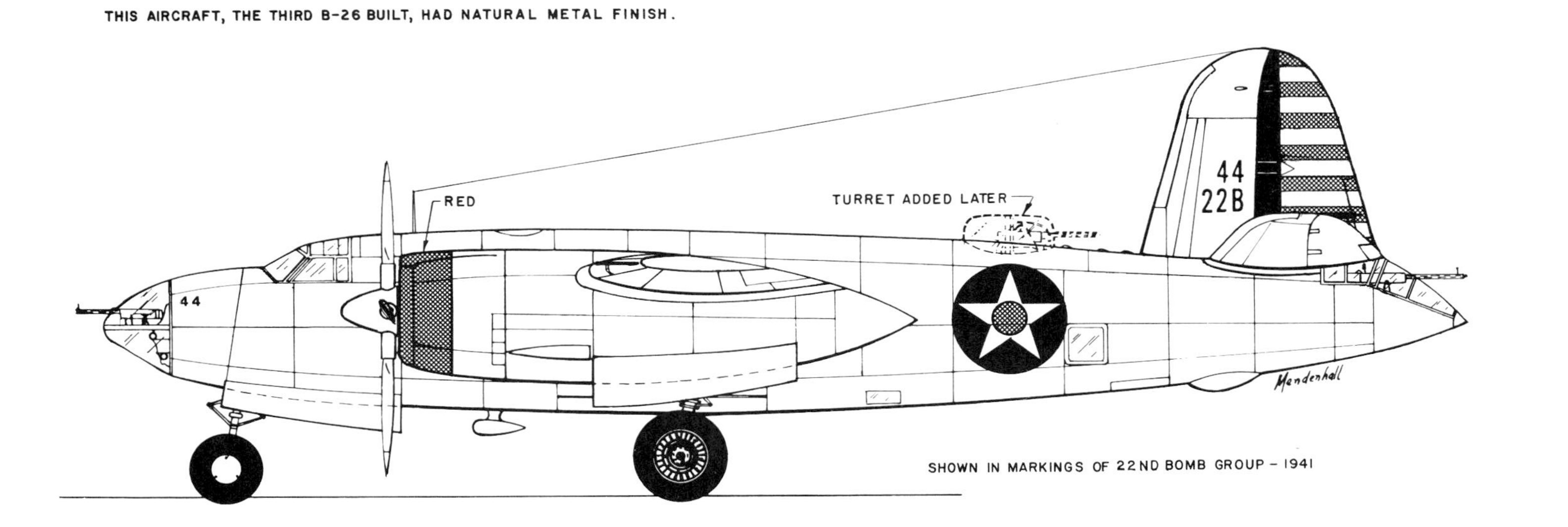

SHOWN IN MARKINGS OF 22ND BOMB GROUP - 1941

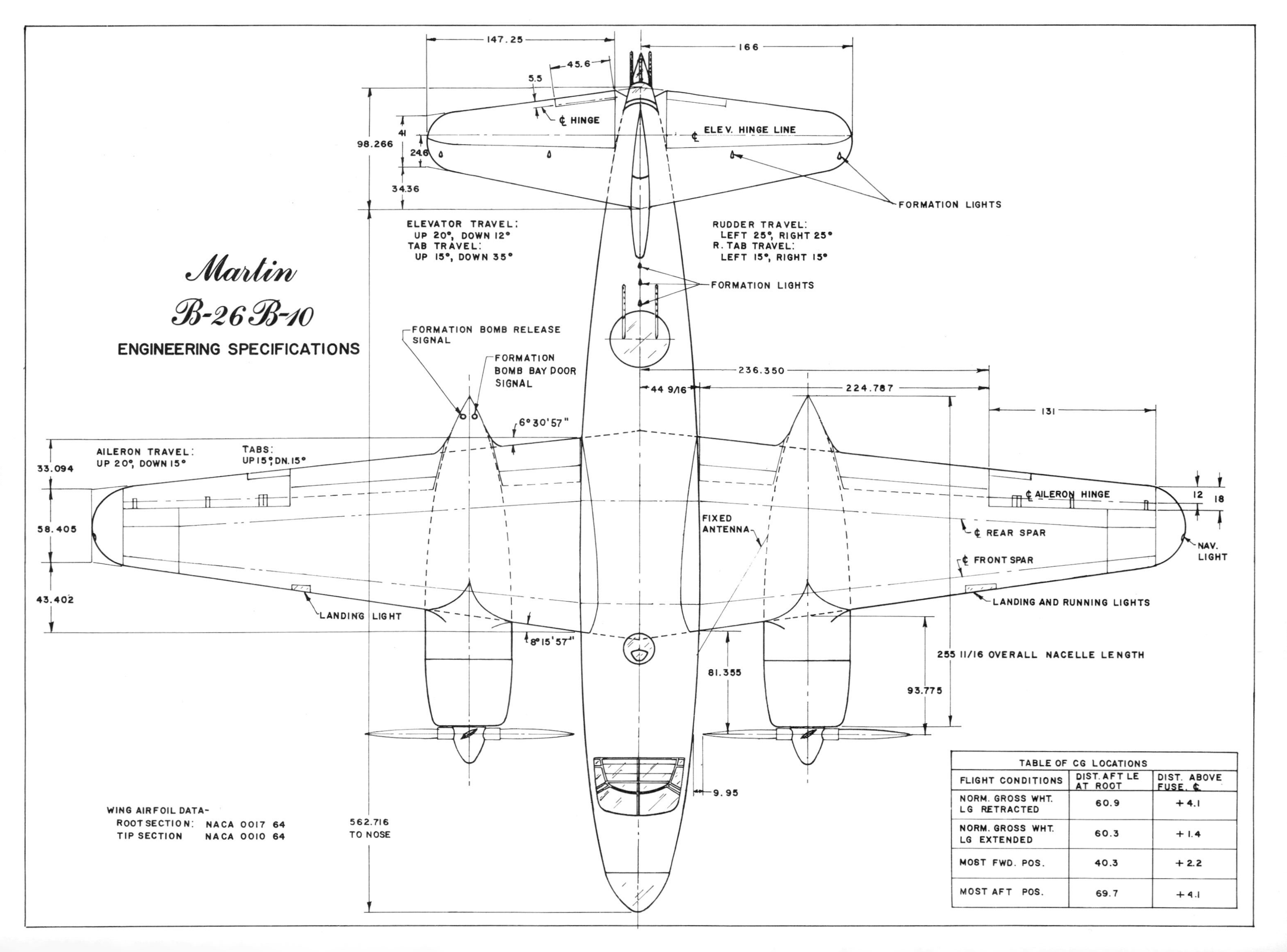

TABLE OF CG LOCATIONS

FLIGHT CONDITIONS	DIST. AFT LE AT ROOT	DIST. ABOVE FUSE. ℄
NORM. GROSS WHT. LG RETRACTED	60.9	+4.1
NORM. GROSS WHT. LG EXTENDED	60.3	+1.4
MOST FWD. POS.	40.3	+2.2
MOST AFT POS.	69.7	+4.1

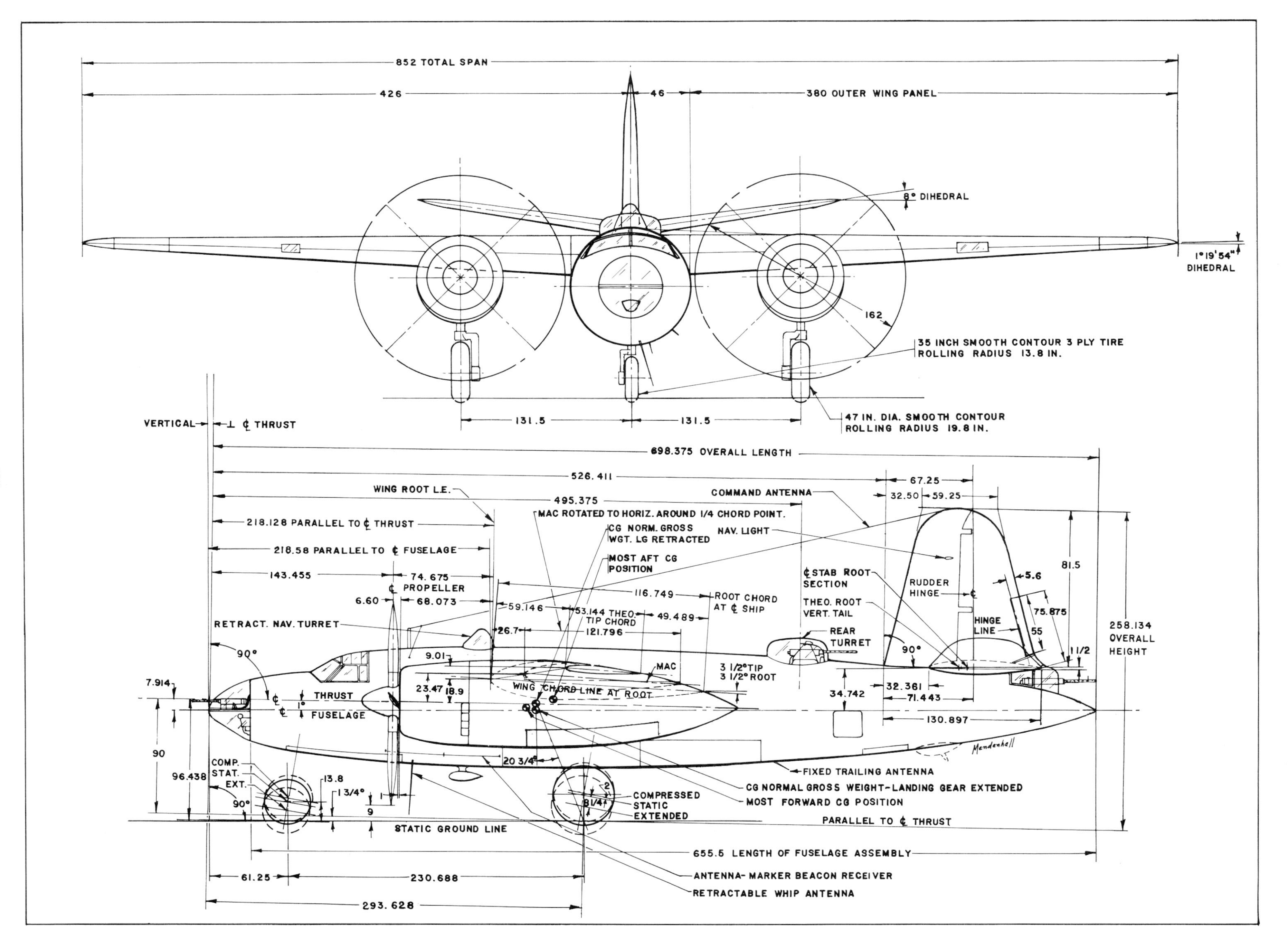
852 TOTAL SPAN
426
46
380 OUTER WING PANEL
8° DIHEDRAL
1°19'54" DIHEDRAL
162
35 INCH SMOOTH CONTOUR 3 PLY TIRE
ROLLING RADIUS 13.8 IN.
47 IN. DIA. SMOOTH CONTOUR
ROLLING RADIUS 19.8 IN.
131.5
131.5
VERTICAL
⊥ ℄ THRUST
698.375 OVERALL LENGTH
526.411
67.25
WING ROOT L.E.
COMMAND ANTENNA
495.375
32.50
59.25
MAC ROTATED TO HORIZ. AROUND 1/4 CHORD POINT.
218.128 PARALLEL TO ℄ THRUST
CG NORM. GROSS
WGT. LG RETRACTED
NAV. LIGHT
218.58 PARALLEL TO ℄ FUSELAGE
MOST AFT CG
POSITION
143.455
74.675
℄ PROPELLER
116.749
ROOT CHORD
AT ℄ SHIP
6.60
68.073
59.146
53.144 THEO.
TIP CHORD
49.489
℄ STAB ROOT
SECTION
RUDDER
HINGE
THEO. ROOT
VERT. TAIL
81.5
5.6
75.875
HINGE
LINE
55
258.134
OVERALL
HEIGHT
RETRACT. NAV. TURRET
26.7
121.796
REAR
TURRET
90°
11/2
9.01
MAC
3 1/2° TIP
3 1/2° ROOT
90°
7.914
23.47
18.9
WING CHORD LINE AT ROOT
32.361
34.742
71.443
℄ THRUST
1°
℄ FUSELAGE
130.897
90
Mendenhall
20 3/4°
96.438
COMP.
STAT.
EXT.
13.8
13/4°
2
COMPRESSED
STATIC
EXTENDED
81/4
1
90°
9
FIXED TRAILING ANTENNA
CG NORMAL GROSS WEIGHT-LANDING GEAR EXTENDED
MOST FORWARD CG POSITION
PARALLEL TO ℄ THRUST
STATIC GROUND LINE
655.5 LENGTH OF FUSELAGE ASSEMBLY
61.25
230.688
ANTENNA-MARKER BEACON RECEIVER
RETRACTABLE WHIP ANTENNA
293.628

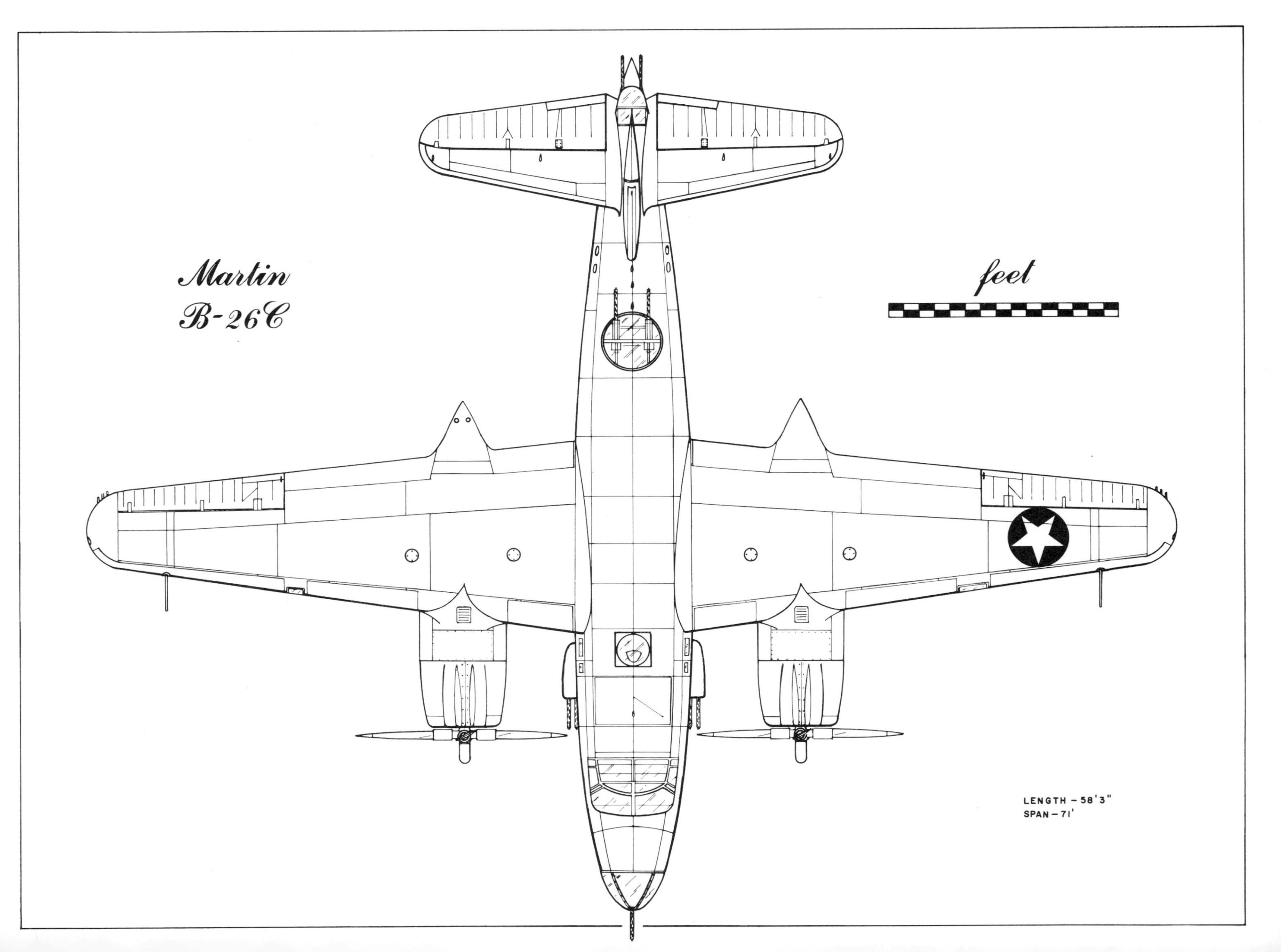
Martin
B-26C
feet
LENGTH – 58'3"
SPAN – 71'

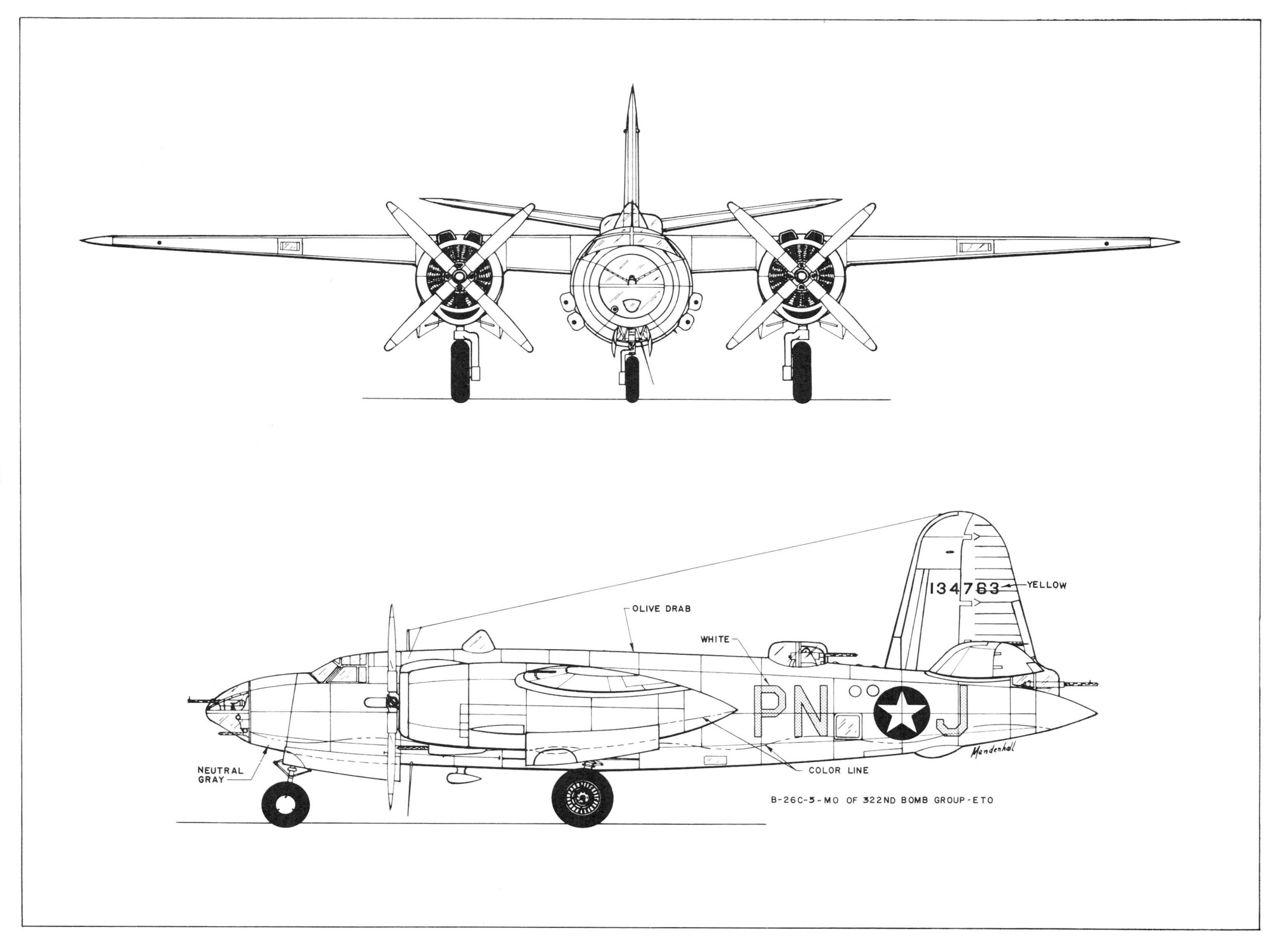

B-26C-5-MO OF 322ND BOMB GROUP-ETO

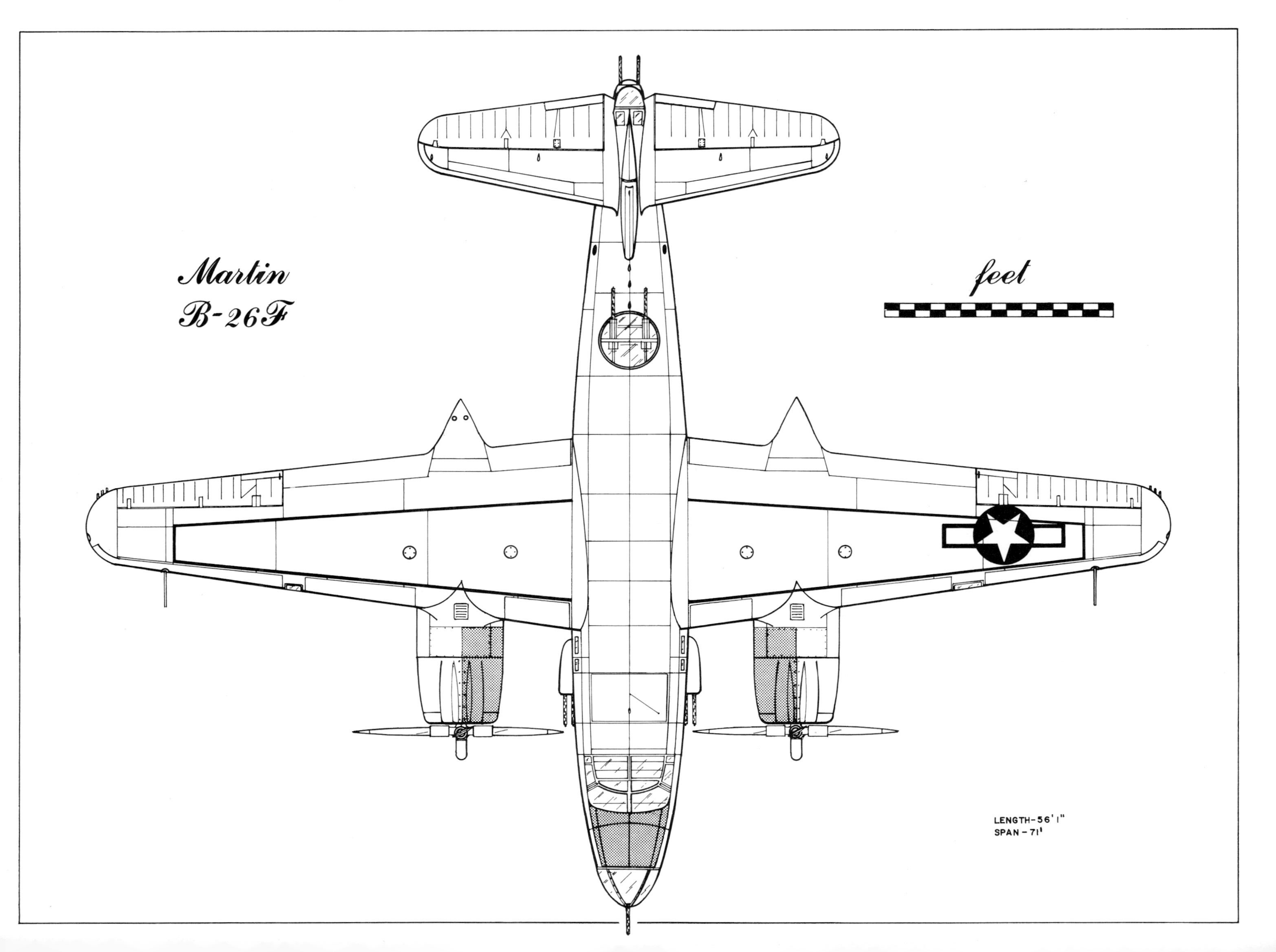
Martin
B-26F
feet
LENGTH-56'1"
SPAN-71'

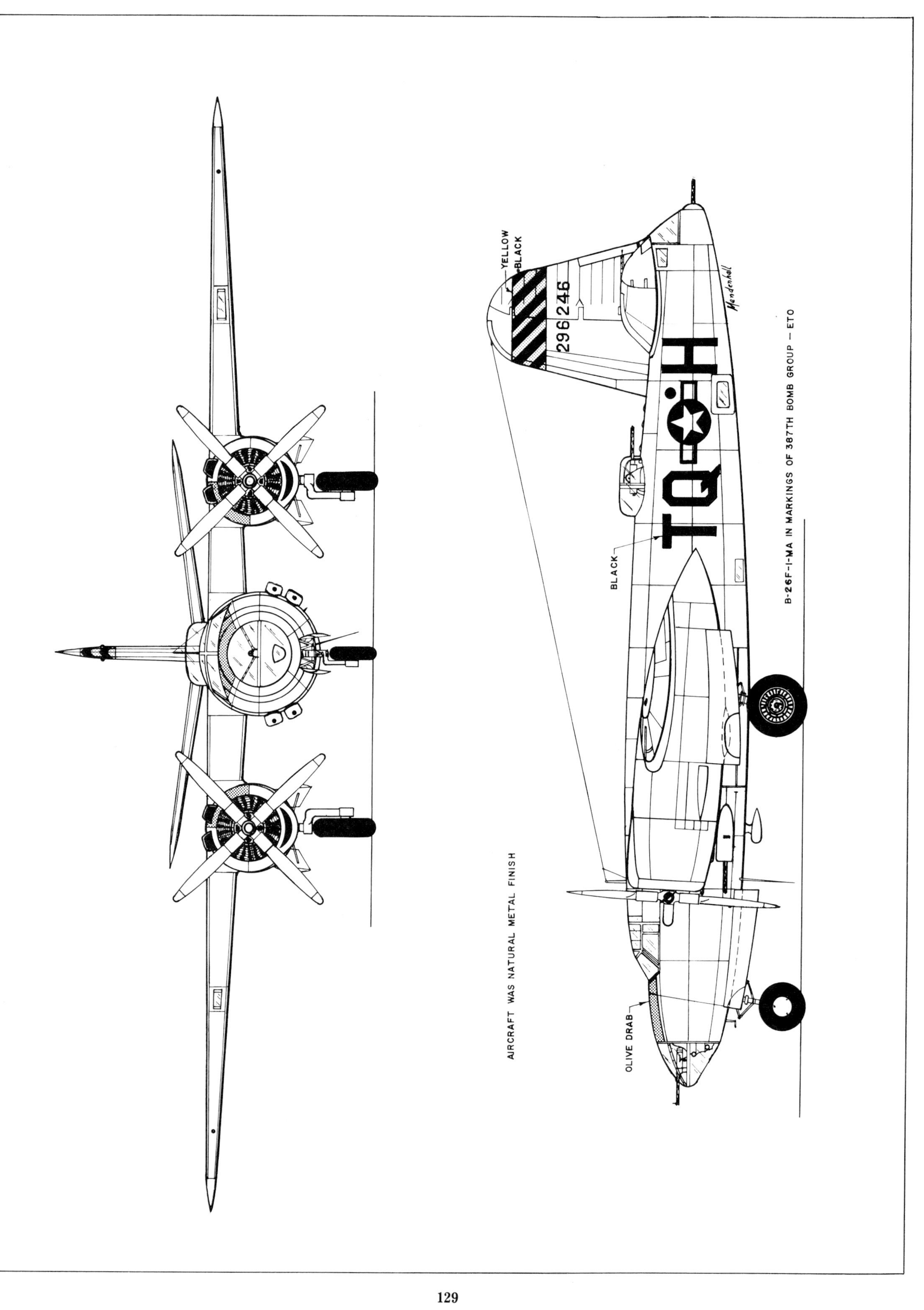

B-26F-1-MA IN MARKINGS OF 387TH BOMB GROUP – ETO

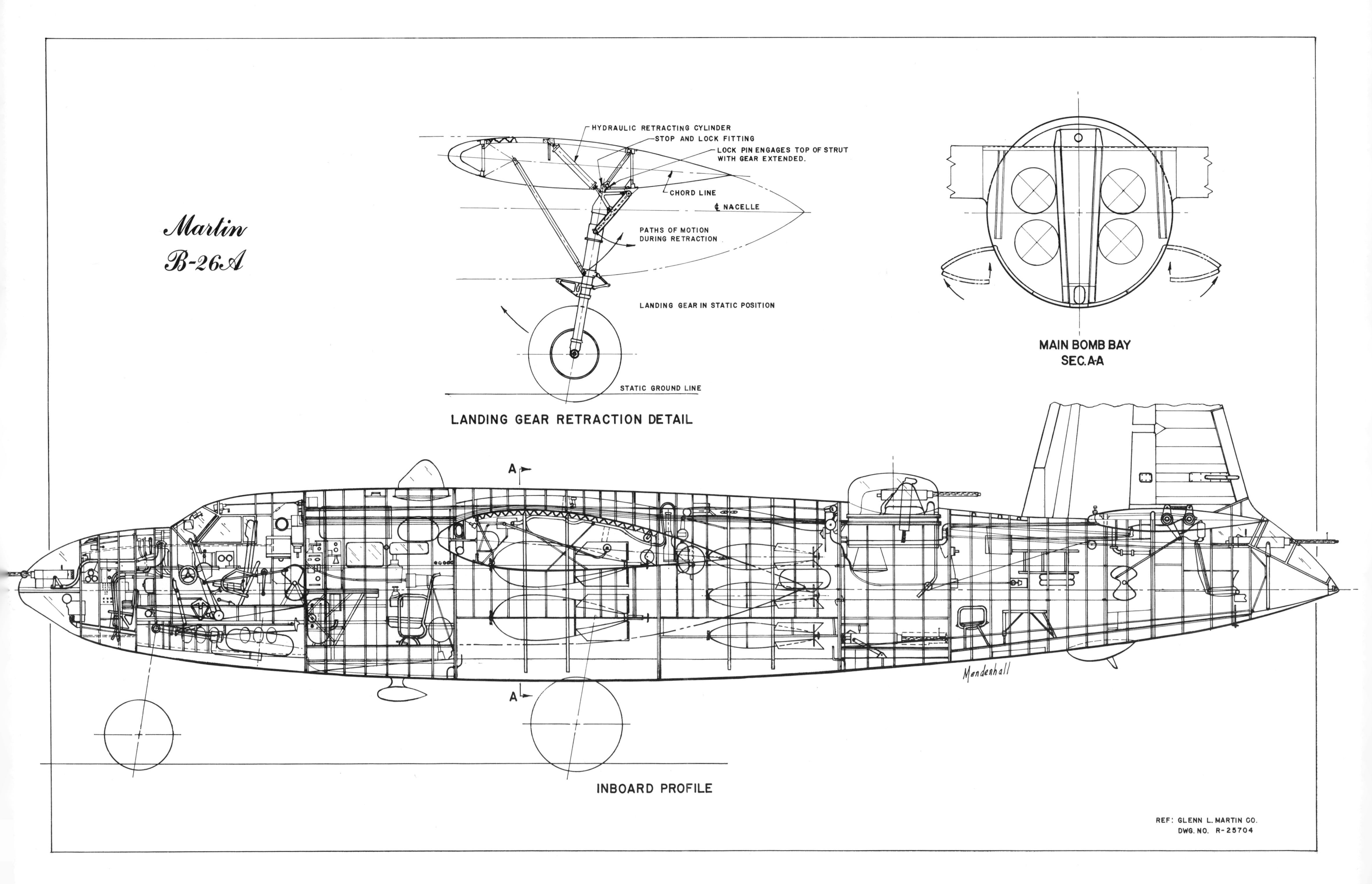
Martin
B-26A
HYDRAULIC RETRACTING CYLINDER
STOP AND LOCK FITTING
LOCK PIN ENGAGES TOP OF STRUT WITH GEAR EXTENDED.
CHORD LINE
℄ NACELLE
PATHS OF MOTION DURING RETRACTION
LANDING GEAR IN STATIC POSITION
STATIC GROUND LINE
LANDING GEAR RETRACTION DETAIL
MAIN BOMB BAY
SEC. A-A
A
A
Mendenhall
INBOARD PROFILE
REF: GLENN L. MARTIN CO.
DWG. NO. R-25704

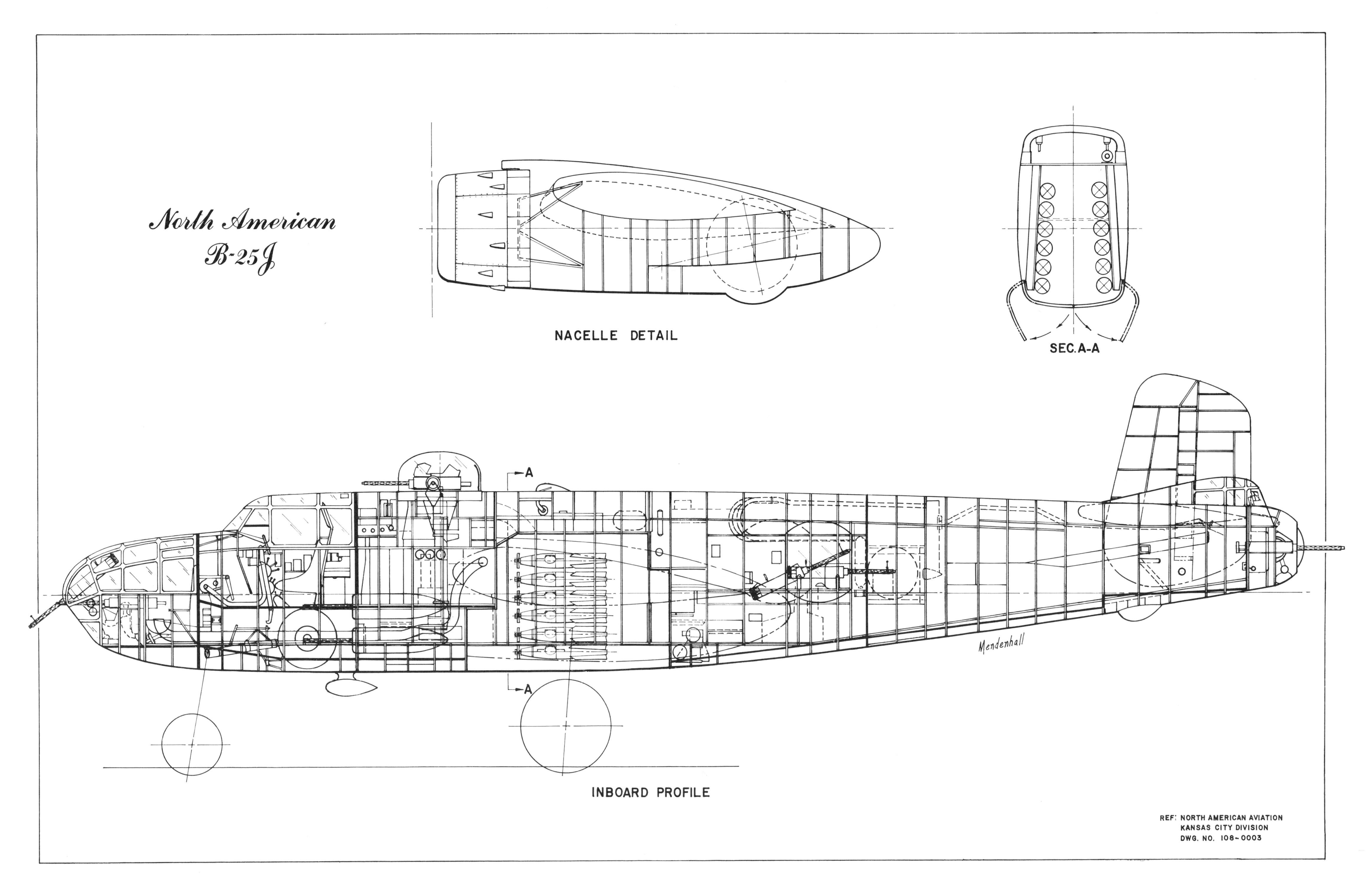
North American
B-25J
NACELLE DETAIL
SEC. A-A
A
A
Mendenhall
INBOARD PROFILE
REF: NORTH AMERICAN AVIATION
KANSAS CITY DIVISION
DWG. NO. 108-0003

9

MEDIUM MAGIC IN RETROSPECT

FROM THE VANTAGE POINT of the 1980's one thing is clear: piston-engined medium bombers started and ended during World War II with the B-25/B-26 aircraft. With some justification certain aero buffs will disagree. What about the other B-26, the Douglas B-26 Invader? Though about the same size, and with greater speed, it came at the wrong time to be considered a true medium in the sense of the aircraft designed to the Wright Field Circular of January 1939. It was not ordered until June 1941 and did not get in the air as a prototype until July 10, 1942. Delivery of service ships to the military did not take place until August 1943. With the upgrading of categories of all bomber sizes it went into combat in September of 1944 as the A-26, a light bomber classification. Remember, by this time the huge B-29 was the heavy bomber entry. With the redesignation of USAF aircraft on June 11, 1948, it finally became the B-26 (the "A" attack designation had been dropped). There were enough of these aircraft, either in service or storage, to fight on with distinction through the entire Korean conflict of the early 1950's. It was a super improved version of the A-20 (DB-7) that had contested the B-25 precursor, the NA-40-2. Had it been around in 1939-40 it would have beat the pants off both the Mitchell and Marauder—but when it did appear it had a few years of the firedrill-paced development of early World War II to help it along.

Of course, the Douglas B-18 derivation of the DC-3 and the more advanced B-23 were also classified as medium bombers, but were never built in any quantity nor used under combat conditions. They were really has-beens that never were.

Pleased with the B-25 and B-26, the Air Corps let fly with another design circular, XC-214, this time for a high altitude medium bomber. That meant cabin pressurization—and Martin and North American were only too eager to oblige with souped-up versions of their standard medium bombers, though the end results didn't look much like either of them. Date of the circular was August 1939—shortly after the headlong engineering dash of both manufacturers to meet the standard medium competition. Martin got XB-27. Its creation looked more like the Grumman F7F-1 Tiger Cat fighter than a derivative of the B-26. The main gear retracted outward into the wings à la Spitfire. The span was 84 feet with a 750-square-foot area. Fuselage length was 60 feet 9 inches and it had a height of 20 feet. Power was to be massive Pratt & Whitney R-2800-9 Twin Wasps providing 2,100 horsepower at takeoff. The service ceiling under these conditions was kicked way upstairs to a stratospheric 33,500 feet. It could carry 4,000 pounds of bombs with a range of 2,900 miles—and the whole thing was dropped after the initial design study. Why? For the same reason as the following North American design effort, the XB-28—it simply wasn't needed.

The North American XB-28 (NA-63) design was an answer to the same circular as the Martin XB-27 and it did better—two copies were actually built. It should have been a Martin design, for it looked much more like the B-26 than the B-25. It was powered by two Pratt & Whitney R-2800-27 Twin Wasps of 2,000 horsepower each. The pressurized fuselage was cylindrical, like the B-26, and 56 feet 5 inches long. Span was 72 feet 7 inches with an area of 676 square feet. This high flying job was to have a service ceiling of 34,600 feet and move along at a top speed of 372 miles per hour; cruise was 255. It could carry 4,000 pounds of bombs. The first prototype, well armed and supercharged to the teeth, had emphasized gun positions and bomb load. The second was more of a high-speed, high-altitude recon plane complete with the latest version of the Fairchild aerial camera line.

The perfecting of low altitude bombing and strafing led the Army to realize that the high altitude medium, while technically great, really had had no use in the war. That was the end of the stratospheric medium bomber requirement. If there *was* a high altitude mission, leave it to the B-17's, B-24's, or the upcoming B-29's.

For the next series of requirements from the Air Force, ongoing technical progress had to be dealt with—the turbojet mainly. With the war still going on, a plethora of other designs with the "B" designation had continued until the North American B-45 Tornado, the first medium-sized jet bomber. As the B-25 to B-28 block of bombers have been covered in detail, a quick run-through of those leading to the B-45 medium bomber is in order:

1. The B-29 was next. This four-engined heavy from Boeing needs no introduction. With its A-bomb drops it spelled the end of World War II.
2. The Lockheed B-30 never happened but it was later to be famous as the Lockheed Constellation transport.
3. The Douglas B-31 was a B-19-sized heavy bomber, four-engined, that just didn't make it off the drawing board. It looked like a super sized A-26 with four engines.
4. The Convair XB-32 was an oversized Liberator and was a backup for the B-29. Its twin tails gave way to a single fin and it went into production and saw limited Pacific service as the Dominator.
5. Martin was back in the high-altitude medium bomber act with the XB-33. Due to predicted performance problems it was abandoned in favor of the XB-33A, a four-engined counterpart of the B-29 and B-32. No one needed a third super heavy and it was canceled.
6. The XB-34 (and B-37) were simply advanced designs of the Navy PV-1 Ventura twin-engine patrol bombers.
7. The Northrop XB-35 was big like a B-36, and a flying wing. Its best performance, of the several built, was in the movie *War of the Worlds.*
8. The B-36 Peacemaker of Consolidated Aircraft needs no introduction. It, of course, was SAC's mainstay of the fifties.
9. The XB-38 and XB-39 were the B-17 and B-29 (with Allison engines).
10. The XB-40 was a B-17 with many guns aboard plus a huge ammunition supply. It was to be used as an escort fighter for long-range missions. It was so heavy it couldn't even keep itself out of trouble with enemy fighters.
11. The XB-41 was a B-24 Liberator turned gunbus and it was even less successful than the B-40.
12. The Douglas XB-42 was just that—an "X". The *Mixmaster* was a last-ditch stand at a layout to wring all there was out of piston engines. Contra-rotating propellers driven by two 1,800-horsepower in-line Allisons propelled the 70-foot span, 54-foot-long craft. Its propellers were on the fuselage tail end.
13. The Douglas XB-43 was the same design warmed over with two GE J-35 turbojets. One copy was built and it introduced the jet engine into the bomber fleet of the USAF.
14. The XB-44 was a B-29 converted to use Pratt & Whitney Wasp Major engines, of 3,000 horsepower each.

15. The North American B-45 Tornado was it. The jet age had arrived! The NA-130 was powered by four Allison J-35 mounted two under each wing. Size was up—96 feet with tip tanks included in the span, length was 75 feet 4 inches. It was bigger, as all things bomberwise were, with the advent of the unlimited horsepower of the jet age. Though larger and more powerful than anything of World War II, the Tornado was the beginning of a medium bomber sized to the jet age. The date of first flight was March 17, 1947.

From that time on came the deluge of designs of jet bombers; the Convair B-46, the Boeing B-47 and the Martin B-48.

It all culminated in the giant Boeing B-52 still flying today as a bulwark against the Red expansion. However, bomber size no longer means anything. A single fighter aircraft with the correct nuclear bomb aboard can drop more explosive power than all the heavy bombers of World War II combined. For a brief four or five year span, however, the piston-engined medium bomber had a use, and served that job well. It was the right stuff at the right time.

10

B25 DATA DIRECTORY

The material presented here is arranged to give a concise, but thorough, source of dimensions, specifications and performance data for the North American B-25 medium bombers. The aircraft having a direct bearing on the B-25's development are covered; for ease of use each major model of the airplane is treated as its own entity.

North American XB-21 Dragon

Number built—1
USAAC serial number—38-485
Charge numbers—NA-21 medium bomber; NA-39 (modifications to NA-21)
Span—95 feet
Wing area—1,120 square feet
Length—61 feet 9 inches
Height—14 feet 9 inches
Powerplants—(2) 1,200 horsepower Pratt & Whitney Hornet R-2180-1 engines with F-10 turbo-superchargers
Empty weight—19,082 pounds
Gross weight—27,253 pounds (maximum overload weight—40,000 pounds)
Armament— (1) .30 caliber machine gun in ball turret nose
(1) .30 caliber machine gun in dorsal turret
(1) .30 caliber machine gun in ventral position
(2) .30 caliber machine guns in two waist positions

North American XB-21 (38-485) (USAF)

Maximum speed—220 miles per hour
Cruising speed—190 miles per hour
Normal bomb load—2,200 pounds
Normal range—1,960 miles with 2,400 gallons of fuel
Crew—6 to 8

This aircraft was designed to be similar to, but better than, the Douglas B-18 Bolo. Service ceiling was 25,000 feet. Five YB-18's were to be used in service trials but the project was canceled. The aircraft was not advanced enough over current production planes to be cost effective.

North American NA-40-1

Number built—1
USAAC serial number—none issued (civil number X-14221)
Charge number—NA-40 twin-engine attack bomber demonstrator
Span—66 feet
Wing area—598.5 square feet
Length—48 feet 3 inches
Height—15 feet 2 inches
Powerplants—(2) 1,100 horsepower 14-cylinder Pratt & Whitney R-1830-S6C3-G Twin Wasp engines
Empty weight—13,000 pounds (approx.)
Gross weight—19,500 pounds
Armament— (4) fixed .30 caliber machine guns in outboard wings
(1) .30 caliber machine gun in nose—500 rounds
(1) .30 caliber machine gun in dorsal turret
(1) .30 caliber machine gun in lower tunnel
Maximum speed—265 miles per hour
Normal bomb load—1,200 pounds
Normal range—1,200 miles
Crew—3

The NA-40-1 was first flown in January 1939 by NAA test pilot Paul Balfour.

North American NA-40-2 (B)

Number built—1 (rework of NA-40-1)
USAAC serial number—none issued (civil number X-14221)
Charge number—NA-40 twin-engine attack bomber demonstrator
Span—66 feet
Wing area—598.5 square feet
Length—48 feet 3 inches
Height—15 feet 2 inches
Powerplants—(2) 1,350 horsepower 14-cylinder Wright Cyclone GR-2600-A71 engines
Empty weight—13,961 pounds
Gross weight—21,000 pounds
Armament— (4) fixed .30 caliber machine guns in outboard wings
(1) .30 caliber machine gun in nose—500 rounds
(1) .30 caliber machine gun in dorsal turret
(1) .30 caliber machine gun in lower tunnel
Maximum speed—285 miles per hour
Normal bomb load—1,200 pounds
Normal range—1,200 miles
Crew—3

The aircraft's service ceiling was 25,000 feet. The single prototype crashed and burned at Wright Field in March 1939. Pilot error, rather than faulty design, was considered to be the reason it was destroyed.

North American NA-40-2 (X-14221) (USAF)

North American B-25

Number built—24
AAF serial numbers—40-2165 through 40-2188
Charge number—NA-62 medium bomber
Span—67 feet 6 inches (67 feet 7 inches after first nine units)
Wing area—610 square feet
Length—54 feet 1 inch
Height—15 feet 9 inches
Powerplants—(2) 1,700 horsepower 14-cylinder Wright Cyclone R-2600-9 engines; 1,350 horsepower at 13,000 feet
Empty weight—16,767 pounds
Gross weight—23,714 pounds
Armament— (1) .30 caliber machine gun with three mounting sockets in nose
(2) .30 caliber machine guns for waist, top and bottom openings
(1) .50 caliber machine gun for prone tail position
Maximum speed—322 miles per hour at 15,000 feet
Normal bomb load—2,400 pounds (maximum overload 3,600 pounds)
Normal range—2,000 miles
Crew—5

The first nine units had straight wings. Number ten, and thereafter, had zero dihedral outer panels giving the gull-wing effect. Approval to build the aircraft was given September 10, 1939. The first flight was August 19, 1940.

North American B-25 (40-2165) (USAF)

North American B-25A

Number built—40
AAF serial number—40-2189 through 40-2228
Charge number—NA-62 medium bomber
Span—67 feet 7 inches
Wing area—610 square feet
Length—54 feet 1 inch
Height—15 feet 9 inches
Powerplants—(2) 1,700 horsepower 14-cylinder Wright Cyclone R-2600-9 engines; 1,350 horsepower at 13,000 feet
Empty weight—16,767 pounds
Gross weight—23,714 pounds
Armament— (1) .30 caliber machine gun with three mounting sockets in nose
(2) .30 caliber machine guns for waist, top and bottom openings
(1) .50 caliber machine gun for prone tail position

North American B-25A
(S/N unknown) (USAF)

North American B-25B
(S/N unknown) (USAF)

Maximum speed—322 miles per hour at 15,000 feet
Normal bomb load—2,400 pounds (maximum overload 3,600 pounds)
Normal range—2,000 miles
Crew—5

Armor plate (3/8-inch) was added to the pilot's, copilot's and bombardier's seats as well as gunner's compartment. The four wing fuel tanks were made self sealing, cutting their capacity from 912 to 694 gallons. A 418-gallon droppable bomb bay tank was designed for long-range ferrying. The service ceiling was 27,000 feet and the aircraft could climb to 15,000 feet in 8.4 minutes. The range with 3,000 pounds of bombs was 1,350 miles. These aircraft were the first to go into operational service with the 17th Bombardment Group. The aircraft appeared in May 1941. By October 1942 two of them, -2196 and -2201, were relegated to training duties as RB-25A's. A few of these were field fitted with external bomb racks for submarine patrol off the U.S. Pacific coast.

North American B-25B (Mitchell I)

Number built—119 (120 built but 40-2243 destroyed before delivery)
AAF serial number—40-2229 through 40-2242
40-2244 through 40-2348
Charge number—NA-62 medium bomber
Span—67 feet 7 inches
Wing area—610 square feet
Length—52 feet 11 inches
Height—15 feet 9 inches
Powerplants—(2) 1,700 horsepower 14-cylinder Wright Cyclone R-2600-9 engines; 1,350 horsepower at 13,000 feet
Empty weight—20,000 pounds
Gross weight—26,208 pounds (28,460 maximum gross)
Armament—(1) .30 caliber machine gun in nose—600 rounds
(2) .50 caliber machine guns in Bendix dorsal turret with 400 rounds per gun
(2) .50 caliber machine guns in periscope sighted Bendix ventral turret with 350 rounds per gun
Maximum speed—300 miles per hour at 15,000 feet
Normal bomb load—2,400 pounds
Normal range—2,000 miles
Crew—5

Twenty-four of these aircraft were reworked for the Doolittle Tokyo raid as follows:

1. Lower Bendix turret removed.
2. Norden bombsight removed and replaced with makeshift low-altitude sight.
3. Wooden dowels added to tail observer's post to simulate machine guns.
4. Gasoline capacity increased to 1,141 gallons with 646 gallons in wings, 225 gallons in bomb bay, 160 gallons in collapsible tank in crawlway, 60 gallons in bottom turret spot plus 10 five-gallon cans for refilling that tank.
5. Takeoff weight was 31,000 pounds
6. Doolittle aircraft was AAF 40-2344, the others were, in order of takeoff: 40-2292, 40-2270, 40-2282, 40-2283, 40-2298, 40-2261, 40-2242, 40-2303, 40-2250, 40-2249, 40-2278, 40-2247, 40-2297, 40-2267, 40-2268.

North American B-25C-NA

Number built—605
AAF serial numbers—41-12434 through 41-13038
Charge number—NA-82 (improved NA-62)
Span—67 feet 7 inches
Wing area—610 square feet
Length—52 feet 11 inches
Height—15 feet 9 inches
Powerplants—(2) 1,700 horsepower 14-cylinder Wright Cyclone R-2600-13 (or -29) engines; Holley carburetors and air filters were fitted
Empty weight—20,000 pounds
Gross weight—29,500 pounds (41,800 maximum gross)

North American B-25C-NA
(41-12800) (USAF)

Armament—(1) .30 caliber machine gun in nose—600 rounds
(2) .50 caliber machine guns in Bendix dorsal turret with 400 rounds per gun
(2) .50 caliber machine guns in periscope sighted Bendix ventral turret with 350 rounds per gun
Maximum speed—295 miles per hour at 15,000 feet
Normal bomb load—2,400 pounds (maximum 3,200 pounds)
Normal range—2,000 miles
Crew—5

The bomb bay racks on the "C" were redesigned from mechanically to electrically operated and to accept the following combinations:
One 2,000-pound bomb
Two 1,600-pound bombs
Three 1,000-pound bombs
Six 500-pound bombs
Eight 250-pound bombs
Twelve 100-pound bombs

A 24 volt electrical system replaced the earlier 12 volt system and anti-icing and deicing systems were added along with self sealing oil tanks. For an aircrew safety improvement, larger escape hatches were provided. After the 383rd aircraft, a scanning blister was added over the navigator's position, and the wing tank fuel capacity was increased to 947 gallons by adding tanks outboard of the existing wing tanks.

North American B-25C-1-NA

Number built—258
AAF serial numbers—41-13039 through 41-13296
Charge number—NA-82 (improved NA-62)
NOTE: For aircraft specifications see parent North American B-25C-NA.

This variant introduced external bomb racks under the wing which held eight 250-pound bombs and short trip bomb loads were increased from 3,200 pounds to 5,200 pounds. Provisions were made for mounting a 2,000 pound torpedo under the fuselage. In that event a bomb bay tank could be used for longer range. The 5th Air Force modified some of their aircraft for strafing missions. The lower turret and bombardier's positions were removed. Four .50-caliber guns were added to the nose plus two .50-caliber package guns on either side. These, combined with the two upper turret guns, totaled ten .50-caliber machine guns that could be brought to bear on a target. Sixty small fragmentation bombs plus six 100-pound demolition bombs could also be dropped to wreck anything the .50's hadn't. A crew of three handled this aircraft.

North American B-25C-5-NA

Number built—162
AAF serial numbers—42-53332 through 42-53493
Charge number—NA-90 Netherlands contract, June 24, 1941
NOTE: For aircraft specifications see parent North American B-25C-NA.

A flexible .50 caliber machine gun replaced the .30 caliber weapon in the nose gun position. It had 300 rounds available. A fixed .50 caliber machine gun was also placed on the right side of the bombardier's position and had 300 rounds available. The USAAF took over this contract from the Netherlands and the aircraft were delivered between October and December 1942.

North American B-25C-10-NA (Mitchell II)

Number built—149 (Actually 150 but one became XB-25E)
AAF serial numbers—42-32233 through 42-32280
42-32282 through 42-32382
Charge number—NA-94 defense aid contract for RAF
NOTE: For aircraft specifications see parent North American B-25C-NA.

One of the aircraft, 42-32281, was pulled from production to rework into the XB-25E-NA, a deicing prototype. The rest of these aircraft had winterization improvements and improved compass equipment over and above previous models.

North American B-25C-15-NA

Number built—145
AAF serial numbers—42-32383; 42-32389 through 42-32532
Charge number—NA-93 defense aid contract for China
NOTE: For aircraft specifications see parent North American B-25C-NA.

Individual exhaust stacks were added for each cylinder to correct long flames from exhaust collector rings that were too visible at night. Five units from charge number NA-93 were completed as B-25G-1-NA.

North American B-25C-20-NA

Number built—200
AAF serial numbers—42-64502 through 42-64701
Charge number—NA-96 (continuation of NA-82)
NOTE: For aircraft specifications see parent North American B-25C-NA.

These aircraft were the same as North American B-25C-15-NA except for a new contract to extend production.

North American B-25C-25-NA

Number built—100
AAF serial numbers—42-64702 through 42-64801
Charge number—NA-96 (continuation of NA-82)
Span—67 feet 7 inches
Wing area—610 square feet
Length—52 feet 11 inches
Height—15 feet 9 inches
Powerplants—(2) 1,700 horsepower 14-cylinder Wright Cyclone R-2600-13 (or -29) engines; Holley carburetors and air filters were fitted
Empty weight—23,000 pounds
Gross weight—33,500 pounds
Armament—(1) .50 caliber machine gun in nose, 400 rounds
(2) .50 caliber machine guns in Bendix dorsal turret with 400 rounds per gun
(2) .50 caliber machine guns in periscope sighted Bendix ventral turret with 350 rounds per gun
Maximum speed—264 miles per hour at sea level; 284 miles per hour at 15,000 feet
Normal bomb load—3,200 pounds
Normal range—1,525 miles
Crew—5

These aircraft completed production of the "C" model at Inglewood, California, in May 1943. There were 230-gallon self-sealing bomb bay tanks available for this model as well as 325-gallon all-metal tanks.

North American B-25D-NC (Mitchell II)

Number built—200
AAF serial numbers—41-29648 through 41-29847
Charge number—NA-87 (Kansas City-built NA-82)
Span—67 feet 7 inches
Wing area—610 square feet
Length—52 feet 11 inches
Height—15 feet 9 inches
Powerplants—(2) 1,700 horsepower 14-cylinder Wright Cyclone R-2600-13 (or -29) engines; Holley carburetors and air filters were fitted
Empty weight—20,000 pounds
Gross weight—29,500 pounds (41,800 maximum gross)
Armament—(1) .30 caliber machine gun in nose—600 rounds
(2) .50 caliber machine guns in Bendix dorsal turret with 400 rounds per gun
(2) .50 caliber machine guns in periscope sighted Bendix ventral turret with 350 rounds per gun
Maximum speed—295 miles per hour at 15,000 feet
Normal bomb load—2,400 pounds (maximum 3,200 pounds)
Normal range—2,000 miles
Crew—5

This aircraft was identical to the Inglewood-built North American B-25C-NA except for its construction in Kansas City. Fisher Body Division of General Motors Corporation was the prime subassembly source for the Kansas City plant, supplying outer wing panels and fuselage side panels. The F-10 reconnaissance aircraft were B-25D-NC's with guns removed and tri-metrogon cameras mounted in the nose for photo-mapping. Ten aircraft were so modified in 1943.

North American B-25D-1-NC

Number built—100
AAF serial numbers—41-29848 through 41-29947
Charge number—NA-87 (Kansas City-built NA-82)
NOTE: For aircraft specifications see parent North American B-25D-NC.

These aircraft, built in Kansas City, picked up the features that were on the B-25-NA after the 383rd aircraft. Main items were adding wing fuel tank capacity, scanning blister over navigator's compartment and carburetor air filters. Like the B-25C-1NA, this aircraft had external bomb racks under the wing capable of carrying eight 250-pound bombs.

North American B-25D-5-NC

Number built—225
AAF serial numbers—41-29948 through 41-30172
Charge number—NA-87 (Kansas City-built NA-82)
NOTE: For aircraft specifications see parent North American B-25D-NC.

Two .50-caliber nose guns were added as on B-25C-5-NA aircraft.

North American B-25D-10-NC

Number built—180
AAF serial numbers—41-30173 through 41-30352
Charge numbers—NA-87 (Kansas City-built NA-82)
NOTE: For aircraft specifications see parent North American B-25D-NC.

As on the B-25C-10-NA, this aircraft had winterization improvements and improved compass equipment.

North American B-25D-10-NC (41-30192) (USAF)

North American B-25D-15-NC

Number built—180
AAF serial numbers—41-30353 through 41-30532
Charge number—NA-87 (Kansas City-built NA-82)
NOTE: For aircraft specifications see parent North American B-25D-NC.

Flame dampening individual exhaust stacks were added to each engine cylinder to lower visibility of exhaust at night. This change was the same as made on B-25C-15-NA.

North American B-25D-20-NC

Number built—340
AAF serial numbers—41-30533 through 41-30847; 42-87113 through 42-87137
Charge number—NA-87 and NA-100 (Kansas City-built NA-82)
NOTE: For aircraft specifications see parent North American B-25D-NC.

A new windshield was added as well as 230-gallon self-sealing bomb bay tanks like those on the B-25C-25-NA. The NA-87 series contract for "D" models was finished in June 1943.

North American B-25D-25-NC

Number built—320
AAF serial numbers—42-87138 through 42-87457
Charge number—NA-100 (continuation of NA-87)
NOTE: For aircraft specifications see parent North American B-25D-NC.

North American B-25D-30-NC

Number built—450
AAF serial numbers—42-87453 through 42-87612; 43-3280 through 43-3479; 43-3480 through 43-3619
Charge number—NA-100 (continuation of NA-87)
NOTE: For aircraft specifications see parent North American B-25D-NC.

North American B-25D-35-NC

Number built—340
AAF serial numbers—43-3620 through 43-3869
Charge number—NA-100 (continuation of NA-87)
Span—67 feet 7 inches
Wing area—610 square feet
Length—52 feet 11 inches
Height—15 feet 9 inches
Powerplants—(2) 1,700 horsepower 14-cylinder Wright Cyclone R-2600-13 (or -29) engines; Holley carburetors and air filters were fitted
Empty weight—23,000 pounds
Gross weight—33,500 pounds
Armament— (1) .50 caliber machine gun in nose, 400 rounds
(2) .50 caliber machine guns in Bendix dorsal turret with 400 rounds per gun
(2) .50 caliber machine guns in periscope sighted Bendix ventral turret with 400 rounds per gun
Maximum speed—264 miles per hour at sea level; 284 miles per hour at 18,000 feet
Normal bomb load—32,000 pounds
Normal range—1,525 miles
Crew—5

These aircraft completed production of the "D" model at Kansas City in March of 1944. Some of the latest B-25D-35-NC's had advanced armament such as was used on the later "H" and "J" models: a tail turret with two .50's plus two .50's in waist positions. In addition to improved winterization changes, some B-25D-35-NC's could add a 125-gallon ferry tank in waist gun position to complement bomb bay ferry tanks.

North American XB-25E-NA

Number built—1
AAF serial number—42-32281
Charge number—NA-94, defense aid contract for RAF
NOTE: For aircraft specifications see parent North American B-25C-10-NA.

This aircraft was originally NA-B-25C-10-NA (42-32281). It was pulled from the production line to install and evaluate a hot-air leading edge deicing proposal.

North American XB-25F-NA

Number built—1
AAF serial number—unknown
Charge number—NA-94, defense aid contract for RAF
NOTE: For aircraft specifications see parent North American B-25C-10-NA.

This aircraft was pulled from the production line to install and evaluate an electric resistance heating type of deicing device.

North American XB-25G-NA

Number built—1
AAF serial number—41-13296
Charge number—NA-82 (improved NA-62)
Span—67 feet 7 inches
Wing area—610 square feet
Length—51 feet
Height—15 feet 9 inches
Powerplants—(2) 1,700 horsepower 14-cylinder Wright Cyclone R-2600-13 engines; Holley carburetors and air filters were fitted
Empty weight—23,000 pounds
Gross weight—35,000 pounds
Armament— (1) nose mounted 75 mm cannon with 21 rounds
(2) .50 caliber fixed nose guns with 400 rounds each
(2) .50 caliber machine guns in Bendix dorsal turret with 400 rounds per gun
(2) .50 caliber machine guns in periscope sighted Bendix ventral turret with 350 rounds per gun
Maximum speed—268 miles per hour at sea level; 281 miles per hour at 15,000 feet
Normal bomb load—3,000 pounds
Normal range—1,525 miles (2,200 miles with ferry tanks)
Crew—5

This aircraft was the last B-25C-1-NA (NA-82) contract unit to be built and the first B-25 to be fitted with a nose cannon for firing trials during flight. The service ceiling was 24,300 feet.

North American B-25G-1-NA

Number built—5
AAF serial numbers—42-32384 through 42-32388
Charge number—NA-93, defense aid contract for China
NOTE: For aircraft specifications see parent North American XB-25G-NA.

These aircraft were the last five B-25C-15-NA's and were converted to 75 mm cannon carriers for service trials. Built under charge number NA-93, these aircraft completed that series.

North American B-25G-1-NA

Number built—100
AAF serial numbers—42-64802 through 42-64901
Charge number—NA-96 (continuation of NA-82)
NOTE: For aircraft specifications see parent North American XB-25G-NA.

These aircraft were originally slated to be B-25C-20's.

North American B-25G-5-NA

Number built—200
AAF serial numbers—42-64902 through 42-65101
Charge number—NA-96 (continuation of NA-82)
NOTE: For aircraft specifications see parent North American XB-25G-NA.

Armor plate (3/8-inch thick) was added to the front and left side of the cockpit, the shell rack, and back protection of the pilot and cannoneer. The bottom periscope sighted turret was dropped after the "G" model aircraft number 221 was manufactured and two of these aircraft were sent to the RAF for trials.

In the Pacific theater the 75 mm cannon was removed on 82 G-5's and replaced with four .50-caliber machine guns in the nose (bringing the total to six). Also, two .30-caliber guns were added to the tail of these machines.

North American B-25G-10-NA

Number built—100
AAF serial numbers—42-65102 through 42-65201
Charge number—NA-96 (continuation of NA-82)
NOTE: For aircraft specifications see parent North American XB-25G-NA.

These aircraft completed the "G" series of cannon carriers, 306 in all. Recognition is easy by noting the twin .50 dorsal Bendix turret is located in the rear fuselage position as on the B-25C & D models.

North American B-25G-10-NA (42-65128) (USAF)

North American B-25H-NA
(Includes B-25H-1-NA through B-25H-10-NA)

Number built—1,000
AAF serial numbers—43-4105 through 43-5104
Charge number—NA-98 (improved development of NA-96)
Span—67 feet 7 inches
Wing area—610 square feet
Length—51 feet
Height—15 feet 9 inches

Powerplants—(2) 1,700 horsepower 14-cylinder Wright Cyclone R-2600-13 engines; Holley carburetors and air filters were fitted
Empty weight—23,000 pounds
Gross weight—35,000 pounds
Armament— (1) Nose mounted 75 mm cannon with 21 rounds
(4) .50 caliber fixed nose guns with 400 rounds each
(4) .50 caliber "package guns" with 400 rounds each
(2) .50 caliber machine guns in Bendix dorsal turret with 400 rounds per gun
(2) .50 caliber machine guns in waist with 200 rounds per gun
(2) .50 caliber machine guns in power operated tail turret with 600 rounds per gun
Maximum speed—268 miles per hour at sea level; 281 miles per hour at 15,000 feet
Normal bomb load—3,000 pounds (or 2,000 pound torpedo)
Normal range—1,525 miles (2,200 with ferry tanks)
Crew—5

The ventral turret was deleted and the Bendix dorsal turret was moved forward to what *was* the navigator's compartment. The 75 mm cannon was the new lighter-weight T13E1 model. Some H's had cannon removed in the field and replaced with two piggyback rockets mounted in that location plus one other slung on the left side of the nose. One H (43-4406) was fitted experimentally with 2,000-horsepower Pratt & Whitney R-2800 engines. The aircraft crashed due to structural failure while on test.

Due to amount of pressure generated in the nose enclosure during firing of four .50's, a vent was placed in the nose to prevent pressurization of that section with the resultant blowing open of the clamshell hood.

The five-man crew on the H was divided in duties as follows:

1. Pilot
2. Navigator, radio operator, cannoneer
3. Flight engineer, dorsal gunner
4. Mid-ship gunner, camera operator
5. Tail gunner

Variants (B-25H-1-NA through B-25H-10-NA) were primarily changes of equipment locations, etc. All aircraft were successive serial numbers within the NA-98 charge number under which 1,000 "G" aircraft were built. The total 75 mm cannon carriers under all charge numbers were 1,306.

North American B-25J-NC (Mitchell III)
(Includes NA-B-25J-1NC through NA-B-25J-35NC)

Number built—4,390
AAF serial numbers—43-3870 through 43-4104 (235 units); 43-27473 through 43-28222 (750 units); 43-35946 through 43-36245 (300 units); 44-28711 through 44-31510 (2,800 units); 44-86692 through 44-86897 (206 units); 45-8801 through 45-8899 (99 units)
Charge number—NA-108, NA-98 with bombardier nose
Span—67 feet 7 inches
Wing area—610 square feet
Length—52 feet 11 inches
Height—16 feet 4-3/16 inches
Powerplants—(2) 1,700 horsepower 14-cylinder Wright Cyclone R-2600-13 (or -29) engines; Holley carburetors and air filters were fitted.
Empty weight—19,530 pounds
Gross weight—26,122 pounds (maximum gross—35,000 pounds)
Armament— (2) .50 caliber machine guns in nose, fixed with 300 rounds per gun
(4) .50 caliber package guns, 400 rounds per gun

North American B-25H-NA (43-4550) (USAF)

North American B-25J-NC (43-3892) (USAF)

(2) .50 caliber waist guns, 250 rounds per gun
(2) .50 caliber machine guns in Bendix upper turret with 400 rounds per gun
(2) .50 caliber machine guns in power-operated tail turret with 600 rounds per gun

Maximum speed—285 miles per hour at 15,000 feet
Normal bomb load—3,000 pounds (or one 2,000-pound bomb)
Normal range—1,525 miles (2,200 miles with ferry tanks)
Crew—6

The B-25J model was produced in far larger quantities than any other B-25 model. Production of this model began in December 1943 and continued until August 1945. Variants generally were minor equipment changes. Some J's were modified to carry eight .50-caliber machine guns in the nose. The glazed nose was replaced with a molded one of the same shape, provided in a kit produced by North American. An extra 150 gallons of fuel was added in the radio compartment of some variants. Four 5-inch rockets could be carried outboard of each nacelle on some models that were fitted with stub rocket racks. Of the 4,390 built, 4,318 were procured by the USAAF and 72 were not delivered. An additional 415 which were on order were canceled.

North American B-25 Special Variants

Besides the mainstream variants just described, there were other designations used to signify specialized uses of the B-25. These were as follows:

CB-25J—These were J's converted after the war for general utility use.

VB-25J—Post war conversions for staff transportation.

AT-24A—Stripped B-25D models, used as advanced trainers.

AT-24B—These were B-25G models used as advanced trainers.

AT-24C—B-25C models used as advanced trainers.

AT-24D—B-25J models used for advanced trainers. (Conversions to AT-24 aircraft totaled sixty.)

TB-25C—These were AT-24C models redesignated in 1947.

TB-25D—AT-24A models redesignated in 1947.

TB-25G—AT-24B models redesignated in 1947.

TB-25J—AT-24D models redesignated in 1947. Additional B-25J conversions to these standards after the war eventually totaled over 600 aircraft.

TB-25K—These were B-25J models modified as E-1 fire control trainers by Hughes in 1951. Total conversions were 117.

TB-25L—Ninety B-25J models were modified to be pilot trainers in 1951 by Hayes.

TB-25M—Forty B-25J models were modified by Hughes as E-5 fire control radar trainers in 1952.

TB-25N—Hayes modified forty-seven B-25J models as pilot trainers in 1954. They were fitted with improved R-2600-29A engines.

F-10—Ten B-25D models were modified by North American in 1943 for photo-mapping. See North American B-25D.

North American TB-25N (0-430716) (USAF)

North American F-10 (RB-25, S/N unknown) (USAF)

North American PBJ Series (U.S. Navy)

The following is a listing of PBJ aircraft:

Model	Quantity	USAAF Model
PBJ-1C	50	B-25C
PBJ-1D	152	B-25D
PBJ-1G	1	B-25G
PBJ-1H	248	B-25H
PBJ-1J	225	B-25J

Some PBJ's had radar in the nose, others on wing tips, for use in night attack fire control. The PBJ-1D models had only a single flexible .50-caliber machine gun in the tail; however, it had .50-caliber waist gun positions as seen on the later B-25H models. The PBJ's frequently carried depth charges along with their bomb loads for anti-submarine warfare.

North American "Mitchell" Series (RAF)

Mitchell I—Twenty-three B-25B's; RAF FK161 through FK183

Mitchell II—Five hundred sixty-nine B-25 C/D's: RAF numbers FL164 through FL218; FL671 through FL709; FL851 through FL874; FR141 through FR209; FR362 through FR384; FR393 through FR397; FV900 through FV999; FW100 through FW280; HD302 through HD345; KL133 through KL161; MA956 through MA957; and two B-25G or Mitchell Series II FR208 through FR209

Mitchell III—Three hundred sixteen B-25J's: RAF numbers HD346 through HD400; KJ561 through KJ800; KP308 through KP328

North American Mitchell III
(Collect Air Photos)

While the United States and Great Britain have been covered in their use of the B-25 (Mitchell) it should be further noted the following countries also made use of the aircraft, but to a somewhat lesser degree. They are arranged in alphabetical order.

Australia (RAAF)

The RAAF operated 50 Mitchells, 39 from Netherlands East Indies stock.

B-25D—A47-1 through A47-25; A47-33 through A47-37

B-25J—A47-26 through A47-32; A47-38 through A47-50

Bolivia (TAM—the military airline)

Thirteen B-25J's were delivered with three converted to military air transports that were still in use in 1979. USAAF serial unknown.

Brazil (Fôrca Aérea Brasileira)

The Brazilian Air Force took delivery of the following B-25's: seven B-25B's (USAAF 40-2245, 40-2255, 40-2263, 40-2306, 40-2309, 40-2310, and 40-2316); twenty-nine B-25C's (USAAF serial numbers unknown); and approximately fifty B-25J's (USAAF serial numbers unknown).

Chile (Fuerza Aérea de Chile)

About one dozen B-25J's were sent to Chile under lend-lease agreements.

China (People's Republic)

The Red Chinese operated a number of captured B-25H's and B-25J's into the early 1960's. NATO code name for these aircraft was "Bank."

China (Taiwan)

The Nationalist Chinese were initially given one hundred thirty-one B-25C's, D's, H's, and J's. During the withdrawal to Formosa (Taiwan) an unknown number were left behind—the People's Republic B-25's.

Columbia

Three B-25's (suffix unknown) were delivered in 1941 and withdrawn from service in 1957. USAAF serials unknown.

Cuba

Some B-25's were delivered in 1947 after signing of Rio Pact. USAAF serial numbers and type are unknown.

Dominica

Two B-25H's were delivered with one sold back to the U.S.A. as N3970C. USAAF serial unknown. Dominican serials were 2501 and 2502.

Ecuador

At least one Ecuadorean B-25 (B-N9069Z) was enlisted in that country's air force. USAAF serial unknown.

Indonesia (Angkatan Udara Republik Indonesia)

About two dozen B-25D's and J's were given Indonesia, on their independence, by the Netherlands.

Mexico

About ten B-25J's were delivered to this country during World War II.

Netherlands East Indies (Militaire Luchtvaart, Koninklijke Nederlandsch Indisch Leger)

Of the 162 B-25C's ordered by the Netherlands East Indies only twenty were built before the NEI fell to the Japanese and they were all allocated to the USAAF. Eventually, however, the NEIAF took delivery of 249 B-25C's, D's, and J's, which generally flew with NEI squadrons attached to the RAAF.

Netherlands Navy (Marine Luchtvaart Dienst)

Twenty-eight RAF Mitchell II's were given the Dutch Government July 22, 1947. They were as follows: ex-RAF FR160, 163, 169, 170, 173, 175, 188, 193, 194, 195, 196, 197, and 199 with Dutch Navy serials 1-11 to 1-23. Three others with Dutch Navy serials 18-1 to 18-3 (ex-RAF FR167, 168, 206) were added. Of the remaining 12 (ex-RAF FR145, 156, 157, 159, 161, 171, 183, 189, 192, 198, 200, and 201) four were serialed 1-24 through 1-27. The rest were used for spare parts.

Peru

Twenty B-25J's were delivered to Peru in 1947. Known USAAF serials are 44-22912; 44-30296; 44-30360; 44-30361; 44-30384; 44-30398; 44-30430; and 44-30418.

Soviet Union (USSR)

Eight hundred seventy B-25B's, C's, D's, H's, and J's were delivered to the Soviet Union under lend-lease. Little is known about their operational use.

Spain (Ejército del Aire)

One B-25D was interned after landing at Melilla, Spanish Morocco in January, 1944. It was used as a staff transport for many years by the Spanish Air Force.

Uruguay

The following eleven B-25J's were delivered to Uruguay: serials G3-150 to G3-160 (USAAF 44-30269; 44-30273; 44-30461; 44-30593; 44-30604; 44-30723; 44-30729; 44-30735; 44-30743; 44-30878 and 44-31190) in 1950. In 1952, three additional B-25J's (G3-161 to -163) and a B-25H (G3-164) were delivered. USAAF serial unknown.

Venezuela

Twenty-four B-25J's were delivered. USAAF serials are unknown. Some of the Venezuelan serial numbers are as follows: 5A40, 6A40, 5B40, 8B40, 15B40, 0953, 1480, 3741, 3898, 4115, 4146, 4173, 5851, and 5880.

North American XB-28 and XB-28A

Number built—1 each
AAF serial numbers—40-3056 (XB-28); 40-3058 (XB-28A)
Charge numbers—NA-63 twin-engine high-altitude bombers; NA-67 (similar to NA-63)
Span—72 feet 7 inches
Wing area—676 square feet
Length—56 feet 5 inches
Height—14 feet
Powerplants—(2) 2,000 horsepower 18-cylinder Pratt & Whitney Twin Wasp R-2800-11 engines
Empty weight—25,575 pounds
Gross weight—37,200 pounds
Armament—(2) .50 caliber machine guns in dorsal turret
(2) .50 caliber machine guns in ventral turret
(2) .50 caliber machine guns in tail barbette
(All the above guns were remote controlled)
Maximum speed—372 miles per hour at 25,000 feet
Normal bomb load—600 pounds
Normal range—2,040 miles
Crew—5

The aircraft cabin was pressurized for high altitude bombing and photography. The service ceiling was 34,600 feet. The XB-28 was fitted out as a bomber; the XB-28A was fitted out for high altitude reconnaissance. A third XB-28 was canceled when the USAAF determined there was no requirement for this type of aircraft.

North American XB-28
(40-3056) (USAF)

11

B26 DATA DIRECTORY

Dimensions, specifications and performance data for the Martin B-26 medium bombers are presented here covering each of the models in chronological order. For ease of use each major model of the airplane is treated as its own entity.

Martin B-26

Number built—201
AAF serial numbers—40-1361 through 40-1561
Constructor's model—179; contract number—AC-13243
Span—65 feet
Wing area—602 square feet
Length—56 feet
Height—19 feet 10 inches
Powerplants—(2) 1,850 horsepower 18-cylinder Pratt & Whitney Twin Wasp R-2800-5 engines
Empty weight—21,375 pounds
Gross weight—27,200 pounds
Armament—(1) .30 caliber flexible machine gun in nose
(2) .50 caliber machine guns in top turret
(1) .30 caliber flexible machine gun in ventral position
(1) .50 caliber flexible machine gun in tail
Maximum speed—315 miles per hour at 15,000 feet
Normal bomb load—4,000 pounds (8,800 pounds maximum)
Normal range—1,000 miles at 265 miles per hour (maximum ferry range was 2,200 miles)
Crew—7

Aircraft was first flown November 25, 1940, by William K. Ebel who was a test pilot and also Martin chief engineer. The landing speed was 103 miles per hour and the takeoff run was 2,500 feet. It required 12.5 minutes to climb to 15,000 feet. The engines produced 1,500 horsepower each at 14,000 feet.

Martin B-26A

Number built—30
AAF serial numbers—41-7345 through 41-7365; 41-7368; 41-7431; 41-7477 through 41-7483
Contract number—AC-13243
NOTE: For aircraft specifications see parent Martin B-26.

The B-26A was the same as the B-26 except for self-sealing fuel tanks, two extra ferry tanks in the rear bomb bay, and 555 pounds of armor plate.

Martin B-26 (40-1503)
(USAF)

Martin B-26A-1 (41-7462)
(USAF)

Martin B-26A-1 (Marauder I)

Number built—109
AAF serial numbers—41-7366 through 41-7367; 41-7369 through 41-7430; 41-7432 through 41-7476
Contract number—AC-13243
NOTE: For aircraft specifications see parent Martin B-26.

Gross weight was 28,376 pounds, empty weight was 21,741 pounds, and the engine was now the Pratt & Whitney R-2800-39. Ferry range was increased to 2,600 miles with 1,462 gallons of gasoline. Some aircraft replaced the .30 caliber gun in the nose with a flexible .50 caliber gun.

Martin B-26B (Marauder IA)

Number built—307
AAF serial numbers—41-17544 through 41-17624; 41-17626 through 41-17851
Contract number—AC-16137
Span—65 feet
Wing area—602 square feet
Length—58 feet 3 inches
Height—19 feet 10 inches
Powerplants—(2) 2,000 horsepower 18-cylinder Pratt & Whitney Twin Wasp R-2800-41 engines
Empty weight—22,380 pounds
Gross weight—29,725 pounds (34,000 pounds maximum gross)
Armament—(1) .50 caliber flexible machine gun in nose
(2) .50 caliber machine guns in dorsal turret
(1) .50 caliber machine gun in ventral position
(2) .50 caliber machine guns in tail, hand-held
Maximum speed—311 miles per hour at 14,500 feet
Normal bomb load—4,000 pounds (8,800 pounds maximum)
Normal range—1,000 miles at 265 miles per hour

The electrical system was changed from 12 to 24 volts. The propeller spinners were deleted and self-sealing fuel lines were installed. A rack was added to the keel to hold a standard 21-inch diameter 2,000-pound torpedo. Nineteen aircraft went to the RAF in the North African theater (FK-362 through FK-380). Internal equipment location was changed for fire extinguishers, water bottles, etc.

Martin B-26B-2-MA

Number built—96
AAF serial numbers—41-17704; 41-17852 through 41-17946
Contract number—AC-16137
NOTE: For aircraft specifications see parent Martin B-26B.

Pratt & Whitney R-2800-41 engines of 1,920 take-off horsepower were installed. The engines provided 1,600 horsepower at 13,500 feet. A 1,500-pound bomb load could be carried 2,000 miles or 4,000-pound load 550 miles. Top speed was 317 miles per hour at 14,500 feet. The service ceiling of the B-26B-2 was 23,500 feet.

Martin B-26B-2-MA (41-17876) (USAF)

Martin B-26B (41-17671) (USAF)

Martin B-26B-4-MA (41-18106) (USAF)

Martin B-26B-3-MA

Number built—28
AAF serial numbers—41-17625; 41-17947 through 41-17973
Contract number—AC-16137
NOTE: For aircraft specifications see parent Martin B-26B.

Powerplants were Pratt & Whitney R-2800-43's and the carburetor intakes were fitted with large cowling airscoops to accommodate tropical sand filters.

Martin B-26B-4-MA

Number built—211
AAF serial numbers—41-17974 through 41-18184
Contract number—AC-16137
NOTE: For aircraft specifications see parent Martin B-26B.

Instrumentation improvements included a life raft that had automatic ejector gear, day and night drift signals, an inclinometer, astro-compass, astrograph, a new starter and winterization equipment. A lengthened nose wheel strut increased the angle of incidence of the wing during takeoff to improve performance in that area. The ventral .30 caliber gun was replaced by two .50 caliber guns in the rear firing through aft fuselage side hatches. Each gun was provided with 240 rounds. The end of the -4 group became -5's and featured mechanically operated wheel well doors and slotted flaps.

Martin B-26B-10-MA (41-18214) (USAF)

Martin B-26B-15-MA (41-31662) (USAF)

Martin B-26B-10-MA (B-26B-1)

Number built—150
AAF serial numbers—41-18185 through 41-18334
Contract number—AC-16137
Span—71 feet
Wing area—659 square feet
Length—58 feet 3 inches
Height—21 feet 6 inches
Powerplants—(2) 1,920 horsepower 18-cylinder Pratt & Whitney Twin Wasp R-2800-43 engines
Empty weight—24,000 pounds
Gross weight—38,200 pounds
Armament—(1) .50 caliber flexible nose gun with 270 rounds
(1) .50 caliber fixed gun in right front fuselage, 200 rounds
(2) .50 caliber package guns, 200-250 rounds
(2) .50 caliber guns in dorsal turret
(2) .50 caliber guns in aft fuselage hatches
(2) .50 caliber guns in tail, hand-held
Maximum speed—282 miles per hour at 15,000 feet
Normal bomb load—4,000 pounds
Normal range—1,000 miles

Takeoff run was reduced from 3,150 feet to 2,850 feet at gross weight of 36,000 pounds due to span being increased to 71 feet with resultant wing area increase of 57 square feet. Addition of rudder and fin size on this variant gave it the familiar name "big assed bird."

Martin B-26B-15-MA

Number built—100
AAF serial numbers—41-31573 through 41-31672
Contract number—DA-46
NOTE: For aircraft specifications see parent Martin B-26B-10-MA.

Martin B-26B-20MA

Number built—100
AAF serial numbers—41-31673 through 41-31772
Constructor's model—179; contract number DA-46
NOTE: For aircraft specifications see parent Martin B-26B-10-MA.

The twin hand-held .50 caliber machine guns in the tail were replaced with a Martin-Bell turret that electro-hydraulically operated the two guns. The gunner was protected by thick glass and an armored bulkhead. Due to the shorter tail turret the overall length of the aircraft decreased to 56 feet 1 inch. Two more 250-gallon fuel tanks were provided for the rear bomb bay.

Martin B-26B-25-MA

Number built—100
AAF serial numbers—41-31773 through 41-31872
Contract number—DA-46
NOTE: For aircraft specifications see parent Martin B-26B-10-MA.

Martin B-26B-30-MA (Marauder II)

Number built—100
AAF serial numbers—41-31873 through 41-31972
Contract number—DA-46
NOTE: For aircraft specifications see parent Martin B-26B-10-MA.

All 100 of these units went to the South African Air Force (FB-418 through FB-517).

Martin B-26B-35-MA

Number built—100
AAF serial numbers—41-31973 through 41-32072
Contract number—DA-46
NOTE: For aircraft specifications see parent Martin B-26B-10-MA.

Martin B-26B-40-MA

Number built—101
AAF serial numbers—42-43260 through 42-43357; 42-43360 through 42-43361; 42-43459
Constructor's model—179; contract number—DA-1049
NOTE: For aircraft specifications see parent Martin B-26B-10-MA.

The rear bomb bay was eliminated on all forthcoming models.

Martin B-26B-40-MA
(42-43459) (USAF)

Martin B-26B-45-MA

Number built—91
AAF serial numbers—42-95738 through 42-95828
Constructor's model—179; contract number—AC-31733
NOTE: For aircraft specifications see parent Martin B-26B-10-MA.

The fixed forward-firing .50 caliber gun was removed on this and all subsequent models.

Martin B-26B-50-MA

Number built—200
AAF serial numbers—42-95829 through 42-96028
Constructor's model—179; contract number—AC-31733
NOTE: For aircraft specifications see parent Martin B-26B-10-MA.

Martin B-26B-55-MA

Number built—200
AAF serial numbers—42-96029 through 42-96228
Contract number—AC-31733
NOTE: For aircraft specifications see parent Martin B-26B-10-MA.

Martin B-26C-MO Series

The B-26C series was the same except for being built in Omaha, as the Martin B-26B-10-MA (long wing) through B-26B-55-MA series. A breakdown by groups follows:

B-26C-5-MO (Includes B-26C-6-MO)
Number built—175
AAF serial number—41-34673 through 41-34847
Contract number—AC-19342
For aircraft specifications see B-26B-10-MA.

B-26C-10-MO
Number built—60
AAF serial number—41-34848 through 41-34907
Contract number—AC-19342
For aircraft specifications see B-26B-10-MA.

Martin B-26C-5-MO
(41-34678) (USAF)

Martin B-26C-15-MO
(41-34911) (USAF)

B-26C-15-MO
Number built—90
AAF serial numbers—41-34908 through 41-34997
Contract number—AC-19342
For aircraft specifications see B-26B-20-MA.
B-26C-20-MO
Number built—175
AAF serial numbers—41-34998 through 41-35172
Contract number—AC-19342
For aircraft specifications see B-26B-20-MA.
B-26C-25-MO
Number built—199
AAF serial numbers—41-35173 through 41-35370; 41-35372
Contract number—AC-19342
For aircraft specifications see B-26B-20-MA.
B-26C-30-MO
Number built—177
AAF serial numbers—41-35374 through 41-35515; 41-35517 through 41-35538; 41-35540; 41-35548 through 41-35551; 41-35553 through 41-35560
Contract number—AC-19342
For aircraft specifications see B-26B-25-MA.
B-26C-45-MO
Number built—334
AAF serial numbers—42-107497 through 42-107830
Contract number—AC-38728
For aircraft specifications see B-26B-55-MA.

Martin B-26C-45-MO (42-107692) (USAF)

Martin B-26D

Number built—1
AAF serial number—unknown
This aircraft was an early (1942) B-26 modified to test hot air deicing of the wing and tail leading edges.

Martin B-26E

Number built—0
This aircraft was a much lightened B-26B with the dorsal turret moved forward to the navigator's compartment—similar to the B-25H model. The weight of this aircraft was to be about 32,000 pounds. None were ever actually constructed.

Martin B-26F-1-MA (Marauder III)

Number built—100
AAF serial numbers—42-96229 through 42-96328
NOTE: For aircraft specifications see parent Martin B-26B-20-MA.
These aircraft, and all F models, had the wing incidence angle increased by 3.5 degrees which gave more propeller ground clearance as well as a more level cruising attitude. The aircraft had no fixed nose gun in bombardier's compartment and the torpedo rack on the fuselage keel was deleted. Fuel capacity was 1,500 gallons.

Martin B-26F-1-MA (42-96316) (USAF)

Martin B-26F-2-MA

Number built—100
AAF serial numbers—42-96329 through 42-96428
NOTE: For aircraft specifications see parent Martin B-26B-20-MA.
One hundred units supplied to RAF and SAAF (HD402-HD501); British equipment was installed on these aircraft.

Martin B-26F-6-MA

Number built—100
AAF serial number—42-96429 through 42-96528
NOTE: For aircraft specifications see parent Martin B-26B-20-MA.
One hundred units were supplied to RAF and SAAF (HD502-HD601) and British equipment was installed on these aircraft.

Martin B-26G-MA Series

This series was the same as the "F" series except AN standard fittings were substituted for AAF fittings to standardize overall military aircraft production. A breakdown by groups follows:
B-26G-1-MA
Number built—100
AAF serial numbers—43-34115 through 43-34214
For aircraft specifications see B-26B-25-MA.

Martin B-26G-1-MA (43-34132) (USAF)

Martin B-26G-10-MA (43-34581) (USAF)

B-26G-5-MA
Number built—200
AAF serial numbers—43-34215 through 43-34414
For aircraft specifications see B-26B-25-MA.
B-26G-10-MA
Number built—125
AAF serial numbers—43-34415 through 43-34464; 43-34540 through 43-34614
For aircraft specifications see B-26B-25-MA.
B-26G-11-MA
Number built—75
AAF serial numbers—43-34465 through 43-34539
For aircraft specifications see B-26B-25-MA.
B-26G-15-MA
Number built—140
AAF serial numbers—44-67805 through 44-67944
For aircraft specifications see B-26B-25-MA.
TB-25G-15-MA
Number built—10
AAF serial numbers—44-67945 through 44-67954
For aircraft specifications see B-26B-25-MA.
B-26G-20-MA
Number built—60
AAF serial numbers—44-67970 through 44-67989; 44-68065 through 44-68104
For aircraft specifications see B-26B-25-MA.
TB-26G-20
Number built—15
AAF serial numbers—44-67955 through 44-67969
For aircraft specifications see B-26B-25-MA.
B-26G-21-MA
Number built—75
AAF serial numbers—44-67990 through 45-68064
For aircraft specifications see B-26B-25-MA.
B-25G-25-MA
Number built—118
AAF serial numbers—44-68105 through 44-68221; 44-68254
For aircraft specifications see B-26B-25-MA.
TB-26G-25-MA
Number built—32
AAF serial numbers—44-68222 through 44-68253
For aircraft specifications see B-26B-25-MA.
These aircraft went to U.S. Navy as JM-2 in March 1945.

Martin XB-26H-MA

Number built—1
AAF serial number—44-68221
NOTE: For aircraft specifications see B-26B-25-MA.
This aircraft was redesignated from B-26G-25-MA to XB-26H for purposes of testing a bicycle landing gear such as was later used on the Boeing B-47 aircraft. The common name for this aircraft was "The Middle River Stump Jumper."

Martin XB-26H-MA (44-68221) (USAF)

Martin AT-23A-MA

Number built—208
AAF serial numbers—42-43358 through 42-43359; 42-43362 through 42-43458; 42-95629 through 42-95737
Contract number—DA-1049 (99 a/c); AC-31733 (109 a/c)
NOTE: For aircraft specifications see B-26B-20-MA.
These aircraft were stripped of combat equipment and used as target towing tugs.

Martin AT-23B-MO

Number built—375

AAF serial numbers—41-35371; 41-35373*; 41-35516*; 41-35539*; 41-35541 through 41-35547*; 41-35552*; 41-35561 through 41-35872*; 42-107471 through 42-107496; 42-107831 through 42-107855

*These models went to U.S. Navy as JM-1's.

Contract number—AC-19342 (324 a/c); AC-38728 (51 a/c)

NOTE: For aircraft specifications see B-26B-20-MA.

These aircraft were stripped of combat equipment and used for target tugs. Three hundred twenty-four of this group became U.S. Navy JM-1's.

Martin AT-23B (41-35872) (USAF)

Martin JM Series (U.S. Navy)

The following is a listing of JM aircraft:

Model	Quantity	USAAF Model
JM-1	194	B-26C-30-MO
JM-2	32	TB-266-25-MA

The tail cone was similar to that used on the original B-26 in the case of the JM series.

Martin "Marauder" Series (RAF)

Marauder I—Fifty-two of the B-26A's RAF numbers FK109 through FK160

Marauder IA—Nineteen B-26B's; RAF numbers FK362 through FK380

Marauder II—One hundred twenty-three B-26C-MO's; RAF numbers FB400 through FB522

Marauder III—Three hundred fifty B-26F and G's; RAF numbers HD402 through HD751

While the United States and Great Britain have been covered in their use of the B-26 (Marauder) it should be further noted that the aircraft also saw service with the French and South African Air Forces.

Free French Air Force

The Free French Air Force received its first Marauders in September 1943 and eventually some 150 B-26C and G models were taken on charge. Examples include:

B-26C-MO—42-107646; 42-107653 through 42-107655; 42-107766; 42-107831 and 42-107894

B-26G-MA—43-34285; 43-34581; 43-34584; 43-34591; 44-67940; and 44-68192

They were operated by two Wings of the 2nd Medium Bomber Brigade, the 31st and 34th.

The 31st Wing was identified by large blue numbers on the vertical fin to denote its groups as follows:

GB I/9 'Gascogne'	51 to 75
GB II/20 'Bretagne'	26 to 50
BG I/22 'Maroc'	01 to 25
GB II/52 'Franche-Comté'	76 to 99

The 34th Wing was identified by large green numbers on the vertical fin to denote its groups as follows:

GB I/32 'Bourgogne'	51-75
GB II/52 'Franche-Comté'	01-25
GB II/63 'Sénégal'	26-50

At war's end all groups, except GB II/20 and GB II/52 were disbanded.

South African Air Force

One hundred Marauder II (RAF serials FB418 through FB517) and numbers of Marauder III's were given by Britain to the South African Air Force equipping Nos. 12, 21, 24, 25 and 30 squadrons in North Africa. Their markings were multi-national—USAAF olive drab and gray, RAF serials, and orange SAAF centers in the RAF roundels.

Martin XB-27

Number built—0

Constructor's model—182

Span—84 feet

Wing area—750 square feet

Length—60 feet 9 inches

Height—20 feet

Powerplants—(2) 2,100 horsepower 18-cylinder Pratt & Whitney Double Wasp R-2800-9 engines

Empty weight—23,125 pounds

Gross weight—32,970 pounds

Armament— (1) .30 caliber machine gun in nose
(1) .30 caliber machine gun in rear cockpit fairing
(1) .30 caliber machine gun that could fire downward or through fuselage sides
(1) .50 caliber gun in tail position

Maximum speed—376 miles per hour

Normal bomb load—4,000 pounds

Normal range—2,900 miles

Crew—7

This aircraft was designed to Air Corps specifications XC-214 for a high-altitude medium bomber. It did not go past the paper planning stage because a need for it had not become apparent. Its service ceiling was estimated to be 33,500 feet.

INDEX

BIBLIOGRAPHY

The American Heritage History of Flight. New York: American Heritage Publishing Co., 1962.

Andrade, John M. *U.S. Military Aircraft Designations and Serials Since 1909.* Leicester: Midland Counties Publications, 1979.

Arnold, Henry H. *Global Mission.* New York: Harper and Brothers, 1949.

Cooper, Bryan. *The Story of the Bomber 1914-1945.* London: Octupus Books Ltd., 1947.

Cornell, James. *The Great International Disaster Book.* New York: Charles Scribner's Sons, 1976.

Fahey, James C. *U.S. Army Aircraft 1908-1946.* New York: Ships and Aircraft, 1946.

Francillon, Renè J. *USAAF Medium Bomber Units, ETO and MTO, 1942-1945.* New York: Sky Books Press Ltd., 1977.

Francis, Devon. *Flak Bait.* New York: Duell, Sloan and Pearce, 1948.

Freeman, Roger A. *The B-26 Marauder At War.* New York: Charles Scribner's Sons, 1978.

Freeman, Roger A. *Camouflage and Markings, Martin B-26 Marauder, No. 14.* London: Ducimus Books Ltd., 1971.

Glines, Carroll V. *Jimmy Doolittle, Daredevil, Aviator and Scientist.* New York: The Macmillan Company, 1972.

Green, William. *Famous Bombers of the Second World War.* Garden City, New York: Hanover House, 1959.

Green, William. *The Observer's World Aircraft Directory.* New York: Frederick Warne and Co., 1961.

Green, William, and John Fricker. *The Air Forces of The World.* Garden City, New York: Hanover House, 1958.

Green, William, and Gerald Pollinger. *The World's Fighting Planes.* Garden City, New York: Hanover House, 1959.

Gurney, Gene. *The War In The Air.* New York: Bonanza Books, 1972.

Jones, Lloyd S. *U.S. Bombers.* Fallbrook, California: Aero Publishers, 1974.

Kohn, Leo J. *Pilots Manual for B-25 Mitchell.* Appleton, Wisconsin: Aviation Publications, 1978.

McDowell, Ernest R. *B-25 Mitchell In Action.* Warren, Michigan: Squadron/Signal Publications, 1978.

McDowell, Ernest R. *North American B-25 A/J Mitchell.* New York: Arco Publishing Co., 1971.

Mesko, Jim. *A-26 Invader in Action.* Carrollton, Texas: Squadron/Signal Publications, 1980.

Misrahi, J. V. *Air Corps.* Northridge, California: Sentry Books, 1970.

Munday, Eric. *USAAF Bomber Units, Pacific 1941-1945.* London: Osprey Publishing Ltd., 1979.

Munson, Kenneth. *Bombers 1939-1945.* New York: The Macmillan Company, 1969.

Swanborough, Gordon. *North American, An Aircraft Album, No. 6.* New York: Arco Publishing Co., 1973.

Taylor, John W. R. *Combat Aircraft of the World.* New York: G. P. Putnam Sons, 1969.

Thetford, Owen. *Aircraft of The Royal Air Force 1918-1958.* London: Putnam, 1958.

Villard, Henry Serrano. *Contact! The Story of the Early Birds.* New York: Bonanza Books, 1968.

Wagner, Ray. *The Martin B-26 B & C Marauder, Profile 112.* London: Profile Publications Ltd. 1971.

Wagner, Ray. *The North American B-25 A to G Mitchell, Profile 59.* London: Profile Publications Ltd., 1966.

Wings and Airpower. Granada Hills, California: monthly 1971 to date.

Martin
B-26A

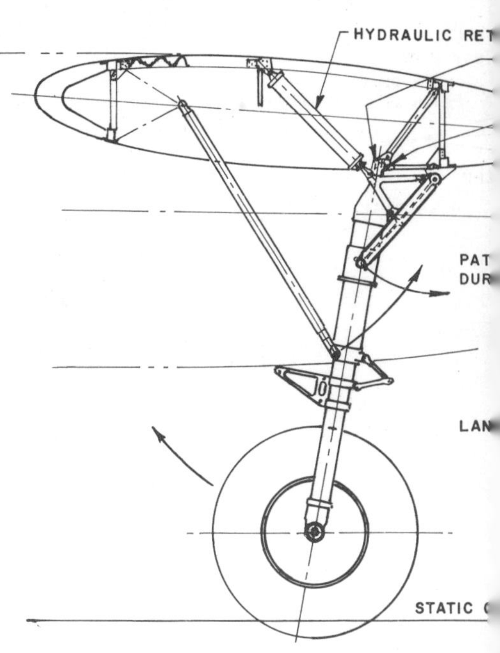

LANDING GEAR RETRACTION D

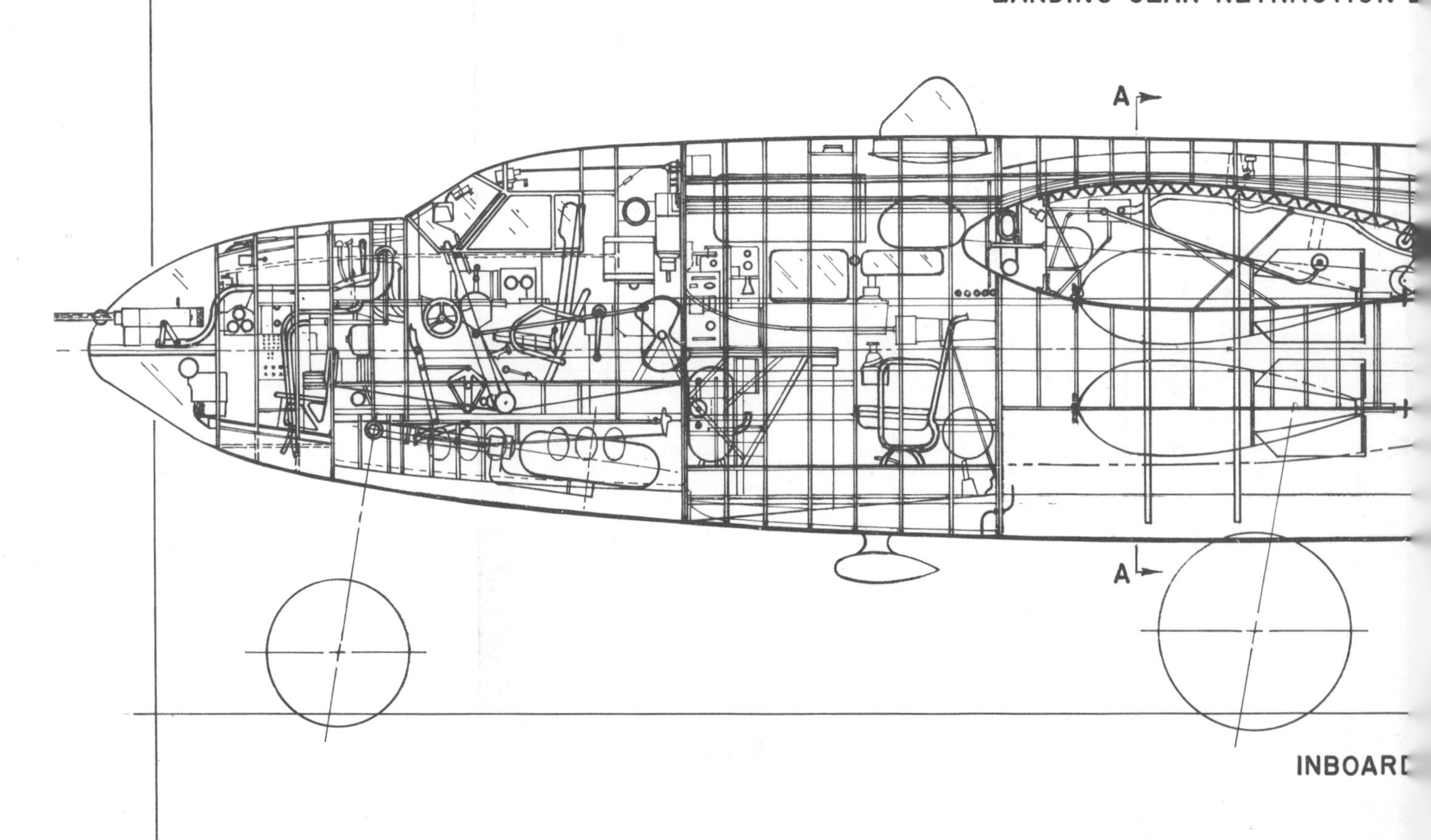

INBOARD